Nursing with a Message

Critical Issues in Health and Medicine

Edited by Rima D. Apple, University of Wisconsin–Madison,
and Janet Golden, Rutgers University, Camden

There is growing criticism of the U.S. healthcare system from consumers, politicians, the media, activists, and healthcare professionals. Critical Issues in Health and Medicine is a collection of books that explores these contemporary dilemmas from a variety of perspectives, among them political, legal, historical, sociological, and comparative, and with attention to crucial dimensions such as race, gender, ethnicity, sexuality, and culture.

For a list of titles in the series, see the last page of the book.

Nursing with a Message

Public Health Demonstration Projects in New York City

Patricia D'Antonio

Rutgers University Press

New Brunswick, New Jersey, and London

Library of Congress Cataloging-in-Publication Data
Names: D'Antonio, Patricia, 1955– author.
Title: Nursing with a message : public health demonstration projects in New York City /
    Patricia D'Antonio.
Description: New Brunswick, New Jersey : Rutgers University Press, 2017. | Series: Critical
    issues in health and medicine | Includes bibliographical references and index.
Identifiers: LCCN 2016015513 | ISBN 9780813571034 (hardcover) | ISBN 9780813571027
    (pbk) | ISBN 9780813571041 (e-book (epub)) | ISBN 9780813571041 (e-book (web pdf))
Subjects: LCSH: Community health nursing—New York (State)—New York. | Public health
    nursing—New York (State)—New York. | BISAC: MEDICAL / Public Health. | MEDICAL
    / Nursing / General. | SOCIAL SCIENCE / Human Services. | MEDICAL / History. |
    SCIENCE / History.
Classification: LCC RT98 .D36 2017 | DDC 610.73/4097471—dc23
LC record available at https://lccn.loc.gov/2016015513

A British Cataloging-in-Publication record for this book is available from the British Library.

Visit our website: http://rutgerspress.rutgers.edu

Manufactured in the United States of America

To Joseph and Frank

Contents

Contents

Illustrations

Acknowledgments

Nursing with a Message was never meant to be a book-length project. Its origins lay in a long-ago request that I present a paper on some aspect of the history of women's health at the 2010 Congress of the International Council on Women's Health Issues hosted by my School of Nursing in Philadelphia. I knew the Barbara Bates Center for the Study of the History of Nursing, my intellectual home for the past thirty years, held Susan Reverby's anthology of pamphlets written by public health nurses involved in an interesting maternal-child health project in the East Harlem section of New York City in the 1920s. I saw this as an opportunity to dive more deeply into these materials. Paper written, paper presented, and then paper put aside.

Until a short time later when my friend and colleague, Julie Fairman, and I visited the Rockefeller Archive Center, successfully persuading the Center to fund a third, but now international, conference on the history of nursing. While there, James Allen Smith—the Center's vice president, its director of Education and Research, and a staunch proponent of the importance of the history of nursing to the Foundation's own history—gave us a tour of its archives. Before me lay boxes upon boxes, row after row, of materials related to the Rockefeller-funded East Harlem Health Demonstration Project and, a short time later, the East Harlem Nursing and Health Demonstration Project whose nurses actually wrote my above referenced pamphlets. Now, I thought, I have access to the materials I needed to return to my paper and turn it into an article-length manuscript. A generous grant-in-aid from the Center allowed me to spend two weeks in Sleepy Hollow, New York, slowly realizing I now had a story that could not be contained by the conventions of a thirty-page manuscript. I thank Jim for believing this project was broader than just the history of nursing—that it was also about the history of public health. And I thank archivist Bethany Antos for constantly steering me in the direction of even more letters and documents.

While in Sleepy Hollow I came to understand the intense optimism that public health reformers of the immediate post–World War I era who believed that they could now put health—health promotion and disease prevention—into public healthcare. This drew me to the collection of the Milbank Memorial Fund that supported its own health demonstration project in what was then the Bellevue-Yorkville section of midtown Manhattan. This has not been a journey of archival hardship. Funding from Penn's own University Research Foundation

(URF) allowed several extended trips to Yale's magnificent Sterling Memorial Library. I admit that my most vivid memory of those visits will remain that of sitting in that appropriately named "cathedral of learning" and feeling the aftershocks of the earthquake that rumbled through the East Coast in 2011.

The URF and the H-15 grant for faculty members from the American Association for the History of Nursing allowed more trips to New York City than I can remember to explore the archives of all the other associations involved with the demonstration projects, particularly those of the Association for Improving the Conditions of the Poor that administered the demonstration projects for both the Rockefeller Foundation and the Milbank Memorial Fund. The staff at Columbia University's Rare Books and Manuscript Library never tired of my repeated requests. I am particularly grateful to Stephen Novak, the head of Archives and Special Collections at the Augustus C. Long Health Sciences Library at Columbia University's Medical Center. He brought to my attention records I would never have found on my own: those of the public health nurses' own Citizen's Protective Society, otherwise known as the Manhattanville project.

There are some very practical debts I need to repay. Most universities, like Penn, have efficient book or article retrieval systems that literally place such requests on your desk or desktop within days. But the staff at Penn Libraries outdid themselves when, three days after I requested that twenty years of the nondigitized journal *Public Health Nursing* be retrieved from storage, two young work-study students arrived at my office with suitcases full of all twenty old and dusty volumes. And as chair of a very large department during the entire life of this story, I would also like to thank my successive assistants, Kristen Nestor, Erica Hildenbrand, and, now, Jake Rutkowski for assiduously protecting the time I had set aside for research and for writing. My thanks also to Lisa Hilmi for her beautiful maps that locate the health clinics and demonstration projects in Manhattan.

I am especially grateful for the community of scholars who have surrounded this story and have helped me develop context, sharpen arguments, and, although I own any remaining errors, save me from some dreadful mistakes. Theodore Brown pushed hard and helped me improve the clarity of some fundamental tensions that ran through the East Harlem Nursing and Health Demonstration Project. Karen Flynn, as always, helped me talk and think through the intersections of race and gender in these demonstration projects, in the larger city, and in the discipline of nursing. David Rosner's, Gerry Oppenheimer's, Daniel Fox's, and James Colgrove's feedback after a seminar at Columbia's Mailman School of Public Health immeasurably strengthened the ending. I am also so thankful to those who have read earlier versions of this story in its entirety: Susan Reverby, Jennifer Gunn, and Cynthia Connolly. Their extraordinarily generous critiques, insights, and advice have made this overall story so much stronger.

The Barbara Bates Center's faculty, staff, and students continuously provided support, collegiality, good council, and friendships. Julie Fairman, Cynthia Connolly, Barbra Mann Wall, Joan Lynaugh, and Jean Whelan have patiently (and sometimes painfully) heard—in formal and informal discussions—about this story since its inception. I owe a special debt to Cynthia Connolly for her strategic advice that constantly moved this story along. Finally, our doctoral students, Kathleen Nishida, Katharine Smith, Amanda Mahoney, and Briana Ralston, sat through a rather muddled presentation of what would become chapter 1. Their advice helped make it chapter 1 and, in return, I hope they learned that historical writing is not a linear process.

I have been very self-conscious during the life of this project that I have been standing on the shoulders of a giant in the history of public health nursing, the late Karen Buhler-Wilkerson. Karen, my dear friend and mentor, set the standard for rigorous scholarship and uproarious fun. I can only aspire to meet her high expectations. And I have always known of my good fortune to sit in a School of Nursing that values history and historical thinking. Both my former dean, Afaf Meleis, and my current dean, Antonia Villarruel, have been unwavering in their support of an intellectual climate that supports the ambitions of the Bates Center faculty and students.

As this story now closes, I thank Janet Golden and Rima Apple, the editors of Rutgers University Press's series on Critical Issues in Health and Medicine, for believing in the importance of this story from the beginning. I also thank Peter Mickulas, senior editor at the Press, for shepherding the story through to publication. As always, my husband, Joseph, and my son, Frank, deserve my eternal gratitude for their patience and pride in my work.

Portions of the following articles are used with the permission of the publishers: "Cultivating Constituencies: The Story of the East Harlem Nursing and Health Service, 1928–1941," *American Journal of Public Health* 103, no. 6 (2013): 988–996; and "Lessons Learned: Nursing and Health Demonstration Projects in New York City, 1920–1935," *Policy, Politics, and Nursing Practice* 14, no. 3–4 (2014): 133–141.

Abbreviations

AICP	Association for Improving the Conditions of the Poor
ARC	American Red Cross
CSS	Community Service Society, Columbia University Rare Book and Manuscript Library
CUMC	Health Sciences Library Archives and Special Collections, Columbia University Medical Center
HSP	Historical Society of Philadelphia
HSS	Henry Street Settlement
LSRM	Laura Spelman Rockefeller Memorial
MCA	Maternity Center Association
MLI	Metropolitan Life Insurance Company
MMF	Milbank Memorial Fund, Sterling Memorial Library, Yale University
RAC	Rockefeller Archive Center
RF	Rockefeller Foundation
TB	Tuberculosis
VNS	Henry Street Settlement and Visiting Nurse Service

Nursing with a Message

Introduction

On March 10, 2010, President Barack Obama signed into law the Patient Protection and Affordable Care Act (ACA). Seven months later a key feature of the bill, the Center for Medicare and Medicaid Innovation (Innovation Center), opened its doors. While the ACA looks to restructure key features of the US healthcare payment systems, the Innovation Center serves as an incubator of new ideas to deliver and pay for care that will improve quality and decrease costs. To this end, its $10 billion budget sets in motion demonstration projects to increase access to high-quality, cost-effective, and coordinated healthcare for beneficiaries of Medicare, Medicaid, and state children's health insurance programs. Its charge is to rigorously and rapidly assess the progress of these demonstrations, and to replicate those with a "high return on investment" in communities across the country. Its first initiative, Strong Start for Mothers and Newborns, has now funded 182 demonstration projects to improve the health of mothers and babies. The intent is that the more successful of these demonstrations can be scaled up to national initiatives that will reduce early elective deliveries, decrease preterm births, test new approaches to prenatal care, and improve outcomes for mothers and babies.[1]

The Center for Medicare and Medicaid Services (CMS) has a thirty-year history of supporting such demonstration projects, most recently in value-based payment systems and disease management and care coordination.[2] Yet demonstration projects in healthcare in the United States predate the CMS's initiatives. *Nursing with a Message* examines the history of the first such demonstration projects in New York City in the 1920s and 1930s, a period commonly referred to as the interwar years. Surprisingly, historians have yet to look systematically at these health demonstration projects that were testing new models of health-care delivery in selected urban and rural communities throughout the country. The brief accounts that do exist are embedded in the histories of the foundations and philanthropies that supported the projects or in the histories of city and state public health departments that looked to them for their policy and practice implications.[3] The East Harlem Nursing and Health Demonstration Project, one of New York City's signature demonstration projects, has had some recognition for its seeming success in settling long-simmering debates about the best orga-nizational structure for public as well as private public health nursing.[4] But this book approaches these demonstrations in New York City as they relate to each other rather than, as in prior work, in isolation.

We need to do this for two reasons. First, there exists an entrenched, yet erroneous, belief that public health prevention and treatment services had their roots in the community health movement of the 1960s. Second, and even more significantly, the United States stands ready to commit significant resources to bolster and expand the capacity of community health centers to provide comprehensive, high-quality, and coordinated care that will target health dis-parities for low-income individuals, racial and ethnic minorities, rural com-munities, and other underserved neighborhoods. It is as committed—as in the past—to identifying and using key quality improvement data to disseminate best practice models to hospitals and healthcare systems throughout the coun-try.[5] It is urgent that we understand the history of an earlier movement also committed to access, quality, care coordination, and data to more fully under-stand all the possibilities and the problems of a national agenda rooted in the needs of particular families and communities.

Three threads, mirroring those of other health demonstration projects throughout the country, ran through all of New York City's projects. The first involved a commitment to broaden public health initiatives to pregnant women and preschool children. These populations had been overlooked in the prewar emphasis on infants and school-aged children. Yet, both mothers and their very young children had, as so many do today, appalling rates of morbidity and mortality. The second centered on initiatives that would teach individuals and families to demand health as well as illness care from their own private physician or, if unable to afford such medical care, from publicly funded clinics. The third was the central place of the public health nurse as the agent who would deliver these messages in her daily rounds in neighbor-hoods and homes. This last thread seemed self-evident. Public health nurses had long considered themselves and had been considered by others as the "connecting link"—between patients and physicians, between and among institutions, and between scientific knowledge and its implementation in the homes they visited. But the nurses in New York City's demonstration projects, like progressive urban colleagues throughout the country, went one step further. They used their experiences in the three demonstration projects to work toward identifying the whole families of their mothers and preschool children as their practice domain.

As historians have long argued, these nurses worked within the interwar years' new constellation of ideas, practices, actions, and actors that shifted the structure of initiatives that might improve the health of women and children in particular, and that of the public in general. The quest to infuse "science" and "scientific meaning" into reasoned, scholarly investigations and also into everyday practices created a new "science of childhood" that emphasized careful developmental studies, on the one hand, and a renewed drive to translate the implications of these results to those individual mothers most responsible for rearing a new generation of upstanding citizens, on the other. Historian Rima Apple's construction of the idea of "scientific motherhood" captures perhaps the strongest impulse to teach mothers the latest science behind such issues as proper prenatal care, infant feeding, and the psychological and environmental requirements to ensure their children's normal growth and development. This impulse, Apple argues, constructed mothers as dependent and passive learners from expert physicians, psychologists, nutritionists, and nurses. But even if believed to be dependent and passive, such mothers were, in fact, quite eager consumers of the literature, lectures, well-baby clinics, and individual conferences that came to large cities and small rural hamlets throughout the country in the 1920s.[6] Most financial support for these came from the unwieldy titled federal legislation, the "Promotion of the Welfare and Hygiene of Maternity and Infancy Act," passed in 1921, that quickly became more popularly known as the "Sheppard-Towner Act" in honor of its legislative sponsors. And while Sheppard-Towner monies did not provide any direct support in New York City's health demonstration projects, it did place nurses in a very direct role in implementing what historians Barbara Beatty, Emily Cahan, and Julie Grant have described as an "empire of child services" created through the 1920s.[7]

This same quest also created a new impulse to, in historian Jodi Vandenberg-Daves's words, "medicalize the maternal body" itself.[8] The timeless debate about whether the process of childbirth required patient watching and waiting as part of a normal experience shared by women across generations or if it needed active and expert intervention by specially trained physicians tipped in the 1920s in favor of skilled medical attendance. Historian Jacqueline Wolfe locates this change in the introduction of obstetrical anesthesia to the birthing process, a change that predated but certainly supported the accelerating 1920s movement to move births from homes into hospitals.[9] And the quest also medicalized—or psychologized—a new framing of how to think about the misbehaviors of children and, especially, adolescents. Historian Kathleen Jones's work on the origins of the "child guidance" movement in the 1920s extended an emphasis on the "whole child" to include his or her emotional as well as physical and developmental life.[10]

Nursing with a Message places New York City's nurses in the middle of this turn toward science. It centers on the power of nurses—too often invisible in histories of healthcare—to also shape the public health messages of the interwar years. These nurses do provide a different lens with which to view this turn: in their day-to-day work with individuals, families, and communities they had to make their own decisions about what aspects of science seemed most relevant, at any one point in time and over the longer time frame within which they envisioned their work. This book draws on Steven Luke's understanding of power as dispositional—that is, it focuses on what these women believed to be their capacity to influence both those they worked for and those they served. The nurses in New York City's health demonstration projects truly occupied a place in the "middle" of the goals of public health reformers, physicians, and patients, and this study shows how they strategically navigated often-rocky shoals. It foregrounds the ideas, the practices, and the effects of the work of these public health nurses as they negotiated their roles within this matrix of competing agendas.

On a broader level, *Nursing with a Message* explores the day-to-day processes involved in the coming together and moving apart of different organizations, disciplinary interests, knowledge domains, and spheres of public and private responsibilities involved in caring for those in need at the point of delivery of service. More specifically, it uses the public health nurses involved in New York City health demonstration projects in the 1920s and the 1930s as a case study of disciplinary tensions inherent in projects with various constituents and invested in multiple and sometimes contradictory outcomes. It shows how one central public health discipline searched for better ways to care for the people it served even as it attended to its own advancement, place, and power in a very complicated space of ideas, practice, action, and actors.

Nursing with a Message centers on three seminal health demonstration projects in New York City in the 1920s and 1930s. Most of its analysis focuses on the East Harlem Nursing and Health Demonstration Project, reconstituted as the East Harlem Nursing and Health Service in 1928. This particular demonstration and later nursing service was completely managed by its public health

Table 1. *Key Features and Funders of the Health Demonstration Projects and Health Clinics in Manhattan*

Demonstration	Purpose	Funder	Dates
East Harlem Health Demonstration Project	*Care Coordination*: To bring all the neighborhood's health and welfare agencies together in one building for "one-stop shopping" Total of 22 health and social welfare agencies involved Each maintained own budget, administrative structures, and client base *Intent*: To "demonstrate" increased service utilization when more accessible; to "demonstrate" feasibility of coordinated neighborhood services; to test premise that physical proximity would eliminate costly service duplication and deliver better health outcomes *Organizational Structure*: Lay health officer presiding over a Community Health Council made up of participating organizations *Goal*: To lay groundwork for a coordinated system of neighborhood health centers that would better integrate the work of private and public health agencies to provide a more seamless experience for individuals and families	Rockefeller Foundation	1921–1932
East Harlem Nursing and Health Demonstration	*Care Control*: Unlike "care coordination" in which individual public and private agencies would maintain control over their own governance and budget, this demonstration in "care control" would pool the personnel and the financial resources of the agencies that provided nursing services to the families of East Harlem—the Henry Street VNS (that provided bedside nursing to the sick in their homes), the AICP (that provided tuberculosis nursing), the Maternal Center Association (that provided prenatal and home birth services), and the Department of Health's nurses (that provided school nursing and well-baby care)—into one controlling organization with its own budget *Intent*: To "demonstrate" the possibility of more efficient use of nursing services; to research the best organization of nursing services; to decrease maternal and infant mortality; to use efficiencies to expand nursing services to preschool children *Organizational Structure*: Independent director of nursing and governing board *Goals*: To lay research groundwork for generalized nursing services as the hallmark of public health nursing practice; to perform service and research	Rockefeller Foundation	1922–1928

(continued)

Table 1. Key Features and Funders of the Health Demonstration Projects and Health Clinics in Manhattan (Continued)

Demonstration	Purpose	Funder	Dates
Bellevue-Yorkville Health Demonstration Project	*Administrative Partnership with the Department of Health* *Intent*: To increase the taxpaying public's willingness to pay for more intensive and educational public health work; to determine best practices in the administration of urban public health work *Organizational Structure*: Led by an officer of the City's Department of Health with supplemental demonstration administrative and clinical staff *Goal*: To lay the groundwork for city-led neighborhood health centers	Milbank Memorial Fund	1926–1932
East Harlem Nursing and Health Service	*Care Control*: (continues) Independent organization *Intent*: To continue a prenatal and preschool child health service; to develop a public health nursing teaching service for postgraduate students *Organizational Structure*: (continues) *Goals*: Service and teaching	Rockefeller Foundation (service) Milbank Memorial Fund (teaching)	1928–1941
Columbus Hill Health Center	*Goal*: Prenatal and infant health teaching *Intent*: To reduce maternal and infant mortality in one poor, black neighborhood *Organizational Structure*: Nurse-managed	AICP	1916–1938

nurses and independent governing board. But it also uses both the Bellevue-Yorkville Health Demonstration Project and, reflecting the city's segregated public health system, the black nurse-managed Columbus Hill Health Clinic, to enlarge, compare, or contrast the ideas and practices developed in East Harlem. Each of these projects had a distinct focus; yet all were linked through the officers of the city's venerable Association for Improving the Conditions of the Poor (AICP), a private charity devoted to providing health and social welfare services to the city's poor and immigrant families, that funneled the philanthropic and foundation monies that made the services possible.

This book is grounded in three central arguments. First, while it is undoubtedly useful to think of these demonstration projects in terms of traditional metrics of successes and failures, such metrics obscure the day-to-day practices and processes involved in turning ideals about health into normative values shared (and performed) by communities. Change did come: New York City's health demonstration projects eventually established what are now the norms for primary, pregnancy, dental, and pediatric care. But, as I argue, it came almost painfully slowly through the day-to-day work of public health nurses going door to door, street to street, school to school, neighborhood to neighborhood, preaching the gospel of good health to those without access to the resources that class, race, ethnicity, and financial stability provided others their messages. Their messages were certainly reinforced by a new group of public health workers called "health educators." But health educators concentrated on crafting messages for groups—of schoolchildren, church members, or club participants. Nurses focused on individuals and families and, conceptually, on those most difficult to reach.

As importantly, change also came through the efforts of families to first incorporate and then normalize these messages of health by removing them from stigmatizing sites of health and social welfare (in which the public health nurses were located) and placing them within the schools that the community embraced. The nurses in New York City's health demonstration projects slowly moved from understanding their role as bringing "medicine and a message" of middle-class values to immigrant families they wished to assimilate, to conceiving of it as being "more than just a messenger" as they sought to be embodiments of a new emphasis on sound mental as well as physical health. Support for public health nursing did decline in the 1930s as nurses painfully realized that it was "not enough to be a messenger." But the decline was less about no longer serving families who needed to assimilate, as other historians have suggested. The decline, I argue, was as much about families taking responsibility for their health and thereby setting limits on the intrusiveness of the

increasingly intimate public health education that came with the public health
turn toward mental health.

I also argue that situating nurses as the focal point of a matrix of competing public health agendas in the interwar year brings into sharper relief the
porousness of professional boundaries in times of intellectual as well as social
change. While traditional histories of public health nursing have highlighted
tensions with physicians, the experiences of nurses in New York City's health
demonstration projects suggest those with female social workers held much
more salience. In *Nursing with a Message* I chart the how the interwar period's
shift of the mental hygiene movement from psychiatry to public health forced
nurses and social workers to rethink both their disciplinary practices and their
relationships with each other. Social workers, not nurses, had developed the
"case work" method for systematically understanding an individual in his or
her environment.[11] But nurses, not social workers, had the experience and the
expertise in the kinds of neighborhood engagement and family outreach necessary for widespread mental health education. What historian Robert Kohler
describes as the war-born enthusiasm for science challenged disciplines,
foundations, and clinicians to rethink norms about what constituted accepted
knowledge and valid evidence.[12] While both nursing and social work drew
on the gendered settlement house traditions of simultaneously incorporating
research and action in their real-world practices, nursing's claim to science—
claims forged in their training school experiences—ultimately strengthened
their place in the increasingly medicalized public health hierarchy.[13] As other
historians have argued, faith in science to find solutions to discrete problems,
the self-proclaimed "new public health" that now focused on the individual
rather than on the environment, as well as the conservative political climate
of the 1920s created a perfect storm that decoupled providing healthcare from
issues of social justice.[14] In ways we have yet to recognize, public health nurses
actively participated in this decoupling process and, I argue, were also central
to the success of this refocused and narrower agenda.

Finally, *Nursing with a Message* argues that history is a valid albeit under-
utilized lens with which to understand current health policy and the processes
of health policy changes. In ways that predate what we now describe as the social
determinants of health, New York's public health leaders, including nurses,
clearly understood the relationships among the conditions in which families
lived, the material resources available to them, the access to education available to their children, and their health status. But issues of access and equity
to the essential health and social services necessary to allow mothers to raise
healthy infants, to help children achieve in school, and to enable breadwinners

to remain productive at work—issues that sound frighteningly similar to those experienced by today's families from vulnerable backgrounds—remained highly problematic.[15] As this story ends in the late 1930s, migrant Puerto Rican and southern black families—experiencing rampant tuberculosis and soaring maternal and infant mortality rates—had moved into the East Harlem neighborhoods. But they also moved into a more medically driven model of public health that nurses actively built. This new public healthcare model did address the healthcare needs of these new constituents. But it also abandoned issues of housing, education, and employment to the more stigmatized domain of social welfare.

The chapters in this book organize these arguments both chronologically and thematically. Chapter 1 maps the social, political, and public health landscape of New York City as it planned to meet the twin challenges of a new health center movement and more effective tuberculosis control and treatment in the aftermath of the First World War. Prominent social workers and physicians found support from the Rockefeller Foundation to create a health center in East Harlem to test the idea that bringing the twenty-three separate agencies that served the neighborhood into one central building could more efficiently coordinate the delivery of health and welfare services to its Italian and Italian American families. These men also found Foundation support for a public health nursing demonstration within a smaller area in East Harlem that would move beyond voluntary care coordination, as would be demonstrated at the health center, to one of care control. All the nurses in the private agencies working in East Harlem would pool their resources, personnel, and dollars into one controlling organization with its own governing board. This particular demonstration would test some of the more vexing issues in the organization and delivery of public health nursing. At the same time, many of these same men found support from the Milbank Memorial Fund for a "monumental enterprise" that included a health center in the Bellevue-Yorkville neighborhood of the city that would provide a model for how to finally eradicate tuberculosis. The city's most prominent public health nurses knew of these plans, and some strongly opposed these ideas. But, in the end, I argue, they felt quite comfortable ignoring them. Leading public health nurses were more concerned about education for practice rather than practice itself.

Chapter 2 delves more deeply into the day-to-day realities of the city's health demonstration projects. It situates these realities amid the tensions between the city's Department of Health and private agencies and associations over who controlled the public health agenda. Both the Rockefeller Foundation and the Milbank Memorial Fund knew that both the private public health nurses working in East Harlem and the city's own public health nurses working in Bellevue-Yorkville were critical to the demonstrations' successes. Indeed, the involvement of the city's own public health nurses working in East Harlem's schools had been a central element of the Rockefeller Foundation's support. The Foundation's policy, both in the United States and abroad, was one of only working through official governmental public health authorities to ensure the sustainability of its initiatives. It hoped to use a consolidated private and public nursing system in East Harlem to ultimately do the same for the city. But, I argue, leading public health nurses shared no interest in this initiative because, in the end, these women won what they had always wanted. By 1928, public health nurses in New York City—not, as in the past, physicians—supervised the independent practices of other public health nurses. They considered this a substantive achievement.

Chapter 3 focuses on the knowledge needed for what contemporaries recognized as "a new approach to health work" among public health nurses. But it is also about how ideas regarding health circulated between and among constituents, how they were implemented, and how their implementation fed back into new policies and practices. It focuses specifically on the complicated and contingent relationships between nurses, social workers, and families at the newly reconstituted East Harlem Nursing and Health Service. Nurses there, like progressive colleagues throughout the country, used their practice experiences to legitimize claims to families as their exclusive domain. They built knowledge that bridged the biological sciences that supported their traditional public health nursing with the new social sciences that buttressed their work with families. This practice, however, brought them out of their traditional disciplinary interests and into a place at the center of their own and also others' agendas. Foundations, families, physicians, and other public health workers all had particular ideas about what nurses should and could do as they delivered their messages of health. As this chapter argues, nurses practiced in a very complicated space of ideas, practice, action, and actors. The knowledge they needed for practice was, in the end, determined not just by the sciences. It was also determined by the demands of the community they sought to serve.

And, as we see in chapter 4, the community around them was changing. The Great Depression had hit East Harlem families early and hard. Its nurses knew about their economic vulnerability, but they thought little of the larger and changing social and healthcare landscape that surrounded them. Through the 1930s Puerto Rican families increasingly settled in neighborhoods of East Harlem. Moreover, these families were moving into a healthcare system increasingly dominated by hospitals and outpatient clinics. I argue that the nurses at East Harlem paid little attention to warnings about the implications of these new clinical sites for healthcare. They steadfastly maintained the site of their practices to that place where it could be most effectively and independently exercised: with cooperative families in their own homes, in the clinics the nurses controlled, and in the classrooms they created. Despite their commitment to maternal-child health initiatives, this narrow focus allowed them to professionally ignore one of the most pressing public health issues in the city in the early 1930s: the newly rising rates of maternal mortality attributed by both the New York Academy of Medicine and the Maternity Center Association to poor obstetrical practices in hospitals that women were increasingly choosing as sites of their infants' births. These nurses could not see or take responsibility for solving problems that lay inside public health policies but outside their defined disciplinary purviews and sites of practice.

As *Nursing with a Message* concludes, it more deeply examines the policy implications we might learn not just from the demonstration projects themselves but also from the work of the nurses who were their public faces. There may be many lessons learned from the East Harlem and Bellevue-Yorkville Demonstration Projects in New York City—lessons such as the need for small, focused projects rather than "monumental" ones, or the need for such projects to have carefully worked through arrangements with all the constituent stakeholders involved in the public's health. But by focusing on the possibilities and the problems that nurses confronted in their day-to-day work with families, we see other lessons. In the end, the nurses in New York City's health demonstration projects did achieve significant successes. They, along with like-minded colleagues, opened public health nursing to interdisciplinary areas of knowledge long before it was popular. They introduced mental health concepts into the practice of nursing long before they became engrained in nursing school curricula. And they broadened their "new approach to health work" to be more inclusive of families rather than individuals.

Yet their history also provides a cautionary message as we move forward to capitalize on the opportunities afforded by the Affordable Care Act and the calls for proposals from the Center for Medicare and Medicaid Innovation. Disciplinary wishes cannot be separated from the needs of constituent communities. The East Harlem Nursing and Health Service ultimately failed because its commitment was to a particular disciplinary mission that emphasized increased educational opportunities for public health nurses. It did meet these nurses' needs. But the service did not meet the needs of the constituent communities it served. From 1928 to 1941, the service focused more on the educational advancement of public health nursing and less on addressing the real needs of constituents in its East Harlem home.

As we look forward to the Center for Medicare and Medicaid's call for demonstration projects like that of the Strong Start for Mothers and Newborns, projects central to nursing's knowledge and practice domains, we can remember the experiences of nurses in East Harlem as lessons about what might be most important. Disciplinary needs—be it East Harlem's role as a teaching center, or now nursing's wish to demonstrate the power of advanced practice nursing, or medicine's wish to lead medical homes—cannot be separated from the needs of constituent communities. These communities might be narrowly defined as the funders of demonstrations or more broadly defined as the people it serves. East Harlem succeeded when it joined with constituents around the need to create meaningful knowledge about how to care for those at home and in the community. It failed when its mission of knowledge generation through research gave way to knowledge transmission through teaching because of a disciplinary commitment to training a new generation of practitioners from across the country and across the globe not shared by those outside its world.

Chapter 1

Medicine and a Message

Public health reformers had every reason for optimism at the dawn of the 1920s. Two seminal events had set grand plans in motion. The first, the decision of the American Red Cross (ARC) that its newly reconfigured peace-time mission would concentrate on the more effective organization of health and social services through neighborhood health centers, promised to solve the knotty problem of care coordination among the myriad of public and private entities operating in large urban areas like New York City. The second, the release of data from the Metropolitan Life Insurance Company's intensive tuberculosis (TB) case finding and treating study in Framingham, Massachusetts, suggested a direct path to bring the "white plague" under control at last.

Yet, New York City's leading public health nurses looked askance at the developing plans to establish the city's own health center and to eradicate tuberculosis—at least as it involved them. They believed they had already solved their discipline's organizational issues with a private system that brought bedside nursing and health teaching to the individual homes of the sick poor and a public system that provided broader communities with health education, immunizations, communicable disease control and quarantines, and the oversight of the health of school-aged children. The city's Henry Street Settlement and Visiting Nurse Service (VNS) was world-renowned for its ability to bring "medicine and a message" of health and American values into the homes of working-class and immigrant families. Its Department of Health, under the tutelage of Lillian Wald, the founder of Henry Street, had the first and now had the largest numbers of nurses working with children in the city's schools.[1]

This chapter maps the social, political, and public health landscape of New York City as it planned to meet these challenges in the aftermath of the First World War. It explores how a small group of white, middle-class, and well-educated public health nursing leaders worked among themselves and with other reformers to consolidate the disciplinary power they gained in their effective work bringing "medicine and a message" of American values to the working, poor, and often immigrant families they served prior to the war. It situates these women within the compromise brokered between public health and private medicine. Bruising battles between public health reformers and representatives of medical practitioners had established firm boundaries regarding who should treat the poor. Those nurses working in public agencies in large urban areas could only teach mothers and children about health and only rarely provided actual home bedside nursing care. In New York City, those working for private agencies like the VNS, the Association for Improving the Conditions of the Poor (AICP), and the Maternity Center Association (MCA) had more latitude. They provided bedside nursing care to sick individuals and prenatal care to mothers even as they taught their families the principles of health and hygiene. They also had a history of strong financial support from the Rockefeller Foundation.

Yet, like their colleagues in other large urban cities, these nurses worked within a complicated matrix that also supported the work of hundreds of other public health nurses employed by small, private neighborhood settlement houses, churches, welfare associations, and community organizations in the city. The proliferation of such agencies across the United States drove the national postwar emphasis on care coordination as a central element of the ARC's commitment to health demonstration projects. In New York City, the problem of so many clinicians working to solve the same kinds of problems brought together the same prominent male social workers and sympathetic physicians to consult with the Rockefeller Foundation and the Milbank Memorial Fund. They successfully found Foundation funding to create a community-based health center in the East Harlem neighborhood of the city that could more efficiently coordinate the delivery of health and social welfare services to those in need; and they dreamed with the Fund's officers of constructing a "monumental enterprise" in the Bellevue-Yorkville districts of the city that would eradicate TB, compel the attention of "scientific men," and force action among communities of voters that seemed far too complacent about the need to increase tax dollars to pay for public healthcare.

The city's leading public health nurses were not invited to these philanthropic tables, although they were aware of the plans. On the one hand, this

omission reeked of the privilege of alliances among powerful white men who
were comfortable in viewing public health nurses as the veritable foot soldiers
of their reform army. But it also kept at a distance those who were not engaged
in their vision. Lillian Wald, the most powerful nursing leader in the city, if not
the world, wanted no part of any planned demonstration either at East Harlem
or Bellevue-Yorkville. East Harlem seemed particularly troubling. In addition
to demonstrating the value of health and social welfare, the second demonstration that involved nurses would be one of care control. The East Harlem
Nursing and Health Demonstration Project intended to pool all neighborhood
nursing personnel and financial resources into one centralized organization to
reduce nursing redundancies and clinical overlaps.

Wald and her public health nursing colleagues centered at the VNS felt
quite comfortable ignoring the plans of other public health reformers. They
believed themselves to be very secure in the putative empire they had built in
New York City, an empire created by well-educated nurses adhering to the highest public health nursing standards when nursing the sick poor in their homes.
But they were well aware that their nurses were an anomaly, not the norm. Wald
and her colleagues were preoccupied with issues surrounding the education for
practice of all public health nurses, not public health practice itself.

Planning for Nursing

Both contemporaries and historians recognized New York City's place at the
epicenter of the public health world in the aftermath of the First World War.
Under the prewar leadership of Hermann M. Biggs, the city attracted international attention for its school health, immunization, tuberculosis, scientific
laboratories, and clean milk reform initiatives. They also recognized the city's
place at the epicenter of the nursing world. Service institutions such as the
VNS at Henry Street and educational initiatives such as those at Teachers College at Columbia University attracted and trained public health nursing leaders
from around the globe.[2]

But for all its successes, postwar New York City still faced seemingly
intractable health issues among its poor, working-class, and immigrant
families—those most vulnerable to the rising costs of living in the postwar city,
labor strikes, and, as the Department of Health reported, the "unstable economic conditions."[3] These health issues included high infant mortality rates,
poor prenatal care, and insufficient attention to the prevention and treatment
of tuberculosis. Established philanthropies, such as the venerable AICP, the
largest and most influential private social service organization in New York
City, provided important financial and social welfare assistance to the city's

own public health initiatives, particularly for families that included a member with tuberculosis. All New York City public health leaders clearly understood the relationships among the conditions in which families lived, the material resources available to them, the access to education available to their children, and their health status. But issues of access and equity to the essential social and health services necessary to allow mothers to raise healthy infants, to help children achieve in school, and to enable breadwinners to remain productive at work remained highly problematic.[4]

The city's nursing leadership, joined by other public health reformers, believed they had another, more vexing, problem to solve in the early 1920s: how would middle-class families who needed care be nursed? In New York City, as in other parts of the country, the working and immigrant poor had access to the services of privately funded visiting nurse services who sent skilled nurses into their homes for short, often daily visits and charged fees that were heavily subsidized by donors. The rich had access to private-duty nurses, graduates of hospital-based training schools, who stayed by their patients' bedsides for the entire illness experience and charged concomitantly higher fees that were beyond the reach of most middle-class Americans. As one commentator noted in 1921, "the great problem" is "the problem of providing adequate nursing service for the community at a rate within the means of those who must pay for such services."[5] Ideas for solving this problem abounded: Have visiting nurse societies engage in the "hourly nursing" of middle-class families at rates greater than those charged the poor but less than the cost of continuous private-duty nursing; have nursing registries—employment agencies that matched a family in need of service with a private-duty nurse in need of work—seek opportunities for nurses who wanted less than continuous employment at prorated fees less than that usually charged; and, to the chagrin of nursing leaders, create a new category of a subsidiary nurse or nurse attendant who had a much shorter period of training.[6]

But in New York City there was cause for some optimism. Nurses Annie Goodrich, who had become head of Henry Street, and Anne Stevens of the Maternity Center Association proposed yet another alternative. They turned to two allies and strong supporters of nursing at the Metropolitan Life Insurance (MLI) Company, Lee Frankel and Louis Dublin. Frankel, the vice president of the company's industrial insurance division, had a long-standing history of collaboration with Lillian Wald at her Henry Street Settlement and Visiting Nurse Service in the early decades of the twentieth century. Wald, known for her innovative approaches to public health nursing, had identified the possibilities of MLI's "penny policies" that—for the penny a week collected door to door, a price within the budget of working-class New Yorkers—policyholders would be eligible for a death benefit that covered funeral expenses. In 1909, Wald had proposed inclusion of an additional benefit. When a policyholder or covered member of his family became ill, Wald would send one of her Henry Street visiting nurses into the home to provide the bedside nursing that could well be life-saving. Dublin was the MLI statistician who proved she was correct. Such nursing both saved lives (and—at 50 cents per visit—supported some of the operating costs of Henry Street) and decreased the dollars in death benefits the company would normally pay. By 1920, such policies had spread like wildfire throughout the country and within the insurance industry itself. Goodrich and Stevens proposed what was essentially a similar, private insurance program, but now for middle-class Americans, that would cover the costs of nursing care.[7]

The proposed Citizen's Health Protective Society's plan would also be much like prevailing mutual aid societies. These societies charged yearly dues and promised families help with medical bills when a member was ill and, most importantly, assistance with funeral expenses if the individual died. Like mutual aid societies, the goal of the Citizen's Health Protective Society was to eventually become a self-reliant, self-governing entity run by its members. But, unlike mutual aid societies, the Citizen's Health Protective Society would help with the costs of health, not illness care, and with the costs of nursing, not medical services. Its ambitious goals were to "work out" a self-supporting nursing service "within the means of the middle class." Concretely, it would provide for the care of pregnant women, assistance at their delivery, and health work with their children until they reached school age. It would also arrange for a visiting nurse to provide bedside nursing when any member became ill. Dues would be $6 each year for an individual and $16 per year for a family.

By 1922, the nurses and their advisors had selected the Manhattanville neighborhood of the city, in the northwest section, from 122nd Street to 142nd Street and from 8th Avenue to the Hudson River because it was a "largely self-supporting neighborhood, not foreign in character and where the vital statistics conform closely to the general average of the city." Manhattanville, in other words, was quite different from the poor, immigrant, and working-class neighborhoods that Henry Street nurses typically served in other Manhattan neighborhoods. It would allow nurses to broaden their reach to a white, middle-class constituency, who lived in newer apartments rather than older tenements, and who were young and newly married and ready to start their families. With the support of an anonymous philanthropist, the new Citizen's Health Protective Society hired its director and set up its office in the heart of the neighborhood. Do you want, it now asked in handouts distributed to the community, a self-supporting nursing and health service?[8]

At the same time, New York City's public health nursing leaders joined others across the United States in seeking answers to what they believed an equally vexing problem: What kind of education did nurses need for public health nursing practice? By the early 1920s, all nurses received their pre-licensure education in hospital-controlled training schools that depended largely on student labor for the care of patients. There, women traded three years of work on the inpatient wards for the knowledge, the clinical opportunities, the diploma they received at graduation, and, if they so chose, the right to sit for state licensing exams and earn the title of "registered nurse."[9]

This training school experience emphasized medical science, skilled techniques, and discipline. Training school experiences varied widely even within New York City. At its worst it meant negligible time in lecture halls, absurdly strict discipline, blind loyalty, and rote obedience. But at its best—and New York City was home to some of the best (albeit segregated) training schools for both white and black nurses in the country—the experience provided the medical knowledge and the training that nurses needed to confront the most persistent challenge to their authority: mothers, drawing on their personal knowledge of their family members in their own homes. Medical knowledge—drawn from the new tenets of exciting developments in bacteriology, microbiology, physiology, and chemistry and learned in a hospital space far from the domestic spaces where they would eventually practice—invited women who would train as nurses to invest themselves with an objective and scientific authority that would more effectively compete with mothers' more personalized and often quite powerful knowledge claims in both the tenements and the drawing rooms of New York City.[10]

Yet, this education and training was for the care of the acutely ill, those recovering from surgeries, trauma victims, birthing mothers, those who required convalescent diets, and, sometimes, sick children. It prepared nurses reasonably well to take care of the sick in their own homes. But it left nurses ill-equipped to do the rest of the work of public health nursing in the early 1920s: to persuade parents to adhere to quarantines if their child had a communicable disease; to monitor the health status of newborn infants at high risk of dying in their first month; to chart the normal development of young children at the baby milk stations where they also received fresh milk; to monitor the status of patients with tuberculosis who lived with their families; and to check for "defects" in the eyes, ears, nose, and throats of school-aged children. Some of New York City's own private and public health nursing agencies had developed their own postgraduate public health nursing training programs for their staff; and a few private and public universities across the United States had begun to develop postgraduate certificate or degree programs to provide classroom content on such topics as methods of organizing and administering public health nursing practices, sanitation, modern social problems, legal and legislative issues, and the knowledge needed for specialized practices in tuberculosis, child welfare, school, and mental hygiene nursing.[11] But, in the eyes of public health nursing leaders, there were too few properly prepared public health nurses and too many, in New York City and across the United States, who held their positions because of rampant political patronage in municipal, county, and state public health systems.[12]

Although preoccupied as a liberal voice in postwar national and international debates over politics, health, and social welfare in an increasingly conservative and nativist United States, Lillian Wald remained the most influential consultant on all matters related to public health nursing in the city and the country.[13] Through her work at Henry Street, Wald cultivated a small group of nursing reformers who shared her vision for both nursing and the health of the community. One of the other leading voices in the campaign to better prepare nurses for public health nursing practice was Annie Goodrich. Goodrich, born into a prominent Connecticut family, had never dreamed of becoming a nurse, but traveled one familiar path into practice. Faced with her family's declining fortune and health, she had entered the New York Hospital's Training School for Nurses in 1890, when Wald was a senior. After graduation she had served as a staunch reform-minded superintendent of several prominent New York City training schools as well as New York State's inspector of nurse training schools, and as a lecturer at Teachers College.

During the First World War, Goodrich and like-minded colleagues orchestrated a major victory for the discipline. As historians have long argued, nursing sick and wounded soldiers had been the only formal way that women could experience war as patriots and citizens.[14] And many American women wanted to serve their country as willing albeit untrained nurses. Recognizing legitimate reports of shortages of trained nurses to care for sick and wounded soldiers—and alarmed by suggestions that the military might turn to well-educated but very quickly trained women volunteer nurses as had England—she campaigned for the establishment of the Army School of Nursing in Washington, DC. The army could meet its shortage by training its own nurses. Goodrich succeeded. And, as the war drew to an end, Goodrich took her place as the inaugural director of the Army School of Nursing in 1918. When the school seemed well established, she returned to Henry Street in 1919 to better manage the day-to-day organization of its VNS.[15]

M. Adelaide Nutting, a music teacher in her native Canada, joined Goodrich in the campaign to reform nursing education. Nutting's path into nursing was another familiar one. Dissatisfied with teaching, she followed other Canadians—drawn by the practice of their famous compatriot Dr. William Osler—to the prestigious Johns Hopkins Hospital Training School for Nurses in Baltimore, Maryland. Nutting had risen to the position of superintendent of nursing and director of its training school by 1895. She had also participated in several seminal events that marked the beginning of the drive to professionalize the discipline: notably, the formation of what would later be renamed the National League for Nursing Education (NLNE) in 1893, and then the American Nurses Association in 1896. She had come to New York City in 1907, holding the first endowed chair in nursing in the country at Columbia University's Teachers College and beginning her long-standing tenure on the board of the Henry Street Settlement and Visiting Nurse Service. And Nutting was fresh from her own World War I victory. She had worked with Vassar College to establish a summer training camp for women college graduates who wanted to contribute to the war as nurses. These women traveled to Poughkeepsie, New York, in the summer of 1918 for an intense immersion in the sciences and public health taught by leading authorities in the field. As the summer closed, these students were sent to participating training schools for the remainder of their clinical experiences.[16]

New York City's nursing leaders also forged strong links with others outside their Henry Street orbit. Lillian Clayton, a 1911 graduate of the nursing program at Teachers College, past president of the NLNE, and current director of the training school at the Philadelphia General Hospital, was one such confidant. Clayton, one of the most respected directors of nurse training schools in the country, found hospital support for moving beyond total reliance on students for all patient care and had hired some graduate-trained nurses. She had begun the process of reshaping class and clinical experiences so that her students had more formal preparation before they entered the hospital's wards. And she worked to develop visiting nursing experiences for some of her interested and talented senior students.[17]

Yet, Mary Beard was the most influential of Wald's circle of nursing reformers. Beard, then the director of Boston's Instructive District Nursing Association, had built her visiting nursing service into one of the largest associations in the country, rivaling only Henry Street in its scope, innovation, and effectiveness. Beard also built ties to the Rockefeller Foundation during her tenure as president of the National Organization for Public Health Nursing. She had been subsequently invited by the Foundation to join Wald, Goodrich, Nutting, and Clayton on the board of directors advising Josephine Goldmark, a progressive labor activist who also lived at Henry Street, in her Foundation-funded inaugural survey of nursing and nursing education in the United States.[18]

This survey, originally commissioned to study the education needed for public health nursing practice by a Foundation with a deep interest in the important role public health nurses played in its own public health and medical education philanthropies, had been planned before the war. Wartime exigencies had forced its postponement. But at a 1920 meeting to discuss reviving the plans, Foundation officials heard Nutting's plea for an enlarged scope of the study. Nutting wanted nothing less than "a serious and thorough study of the entire system of nursing education." They also heard from Herbert Mills, a Vassar professor deeply involved with construction and implementation of the summer training camp, who spoke in support of a broader scope. His own college graduates, he told Foundation officials, complained "bitterly of hard work and long hours" when they left the camp and entered training schools.[19] A consensus emerged rather quickly. There would now be two reports contained in the one formal survey: the first on public health nursing, in particular, and the second on pre-licensure nursing education, in general. American nursing leaders awaited the report, due in 1923, with baited breath. They anticipated the report would do for nursing education what the Carnegie Foundation–funded report on *Medical Education in the United States and Canada* had seemingly done for physicians when it was released in 1910.[20] These nurses hoped this upcoming report would completely transform nursing's educational landscape.

Planning for New York City

At the same time, the American Red Cross had decided that its newly reconfigured peacetime mission would concentrate on the more effective coordination of available social and health services in areas where they already existed; and in the development of new ones in more poorly served parts of the country. It charged local chapters with bringing together community leaders in government, philanthropy, and business to create carefully constructed and coordinated "health centers" that would best serve the needs of defined constituents. In Boston, for example, the city's health department took the lead in establishing the Blossom Street Health Unit for the North End's predominately eastern and southern European immigrant families. With the financial help of philanthropist George Robert White, it brought together the city's private Community Health Association (a new name for its own visiting nurse service), the Family Welfare Society, the Catholic Charitable Bureau, and the Associated Jewish Philanthropies in one building for more effective social service and healthcare coordination. On the other side of the country, in another example, the vast Los Angeles County decentralized its health department and encouraged more rural areas to test different ways of providing easier social service and health-care access to the Mexican and black families they served.[21]

And preliminary data from the Community Health and Tuberculosis Demonstration Study that had begun in 1916 in Framingham, Massachusetts, suggested that early and intensive case finding and treatment decreased overall tuberculosis mortality. The Framingham Study, set in a small community west of Boston, started with all the elements of success. It was a "typical community" of second- and third-generation white Irish Americans, whose immigrant population of 27 percent mirrored that of the United States as a whole. Moreover, it had a supportive and engaged citizenship, a strong public health infrastructure, and a group of private medical practitioners who welcomed the use of the resources and laboratories available through the Study. It had set an ambitious goal: acknowledging the declining rates of TB throughout the United States, it intended to bring death rates to the aspirational rate of 30 per 100,000 at a time when similar communities were experiencing those approximating 121. It brought nurses into the town to canvas its men, women, and children in homes, schools, and workplaces. It offered expert consultative services to local medical practitioners, treatment in one of Massachusetts's TB sanatoria, and, what most families preferred, treatment at home under strict public health nursing supervision. The Study met its goals. But questions remained. Was Framingham truly typical? What role did rising standards of living play in declining TB mortality? And, most important to Bailey Barton Burritt, the general director of the Association for Improving the Conditions of the Poor (AICP) and an influential voice among New York City social reformers, how did the results of the Study support his own idea about the importance of a "home hospital"? Beginning in 1912, the AICP in collaboration with the city's Department of Health had challenged the orthodoxy of hospitalizing those with tuberculosis and, instead, rented an apartment building for a demonstration of the effectiveness of keeping families intact, with decent housing and nutritional support, and under strict public health nursing supervision. The small demonstration collapsed during the war when the city withdrew its funding. But Burritt's dream lived on.[22]

Burritt, educated at Columbia University, was a key member of a small group of New York City's public health reformers who shared interlocking ties of class, gender, race, and progressive vision. He served on the committee advising the nurses' Citizen's Health Protective Society. But he was most active in providing the initial leadership of plans for the city's response to the ARC call and the Framingham Study data. He had been appointed to the general director's position of the AICP in 1914 at a time when Albert Milbank, president of the Milbank Memorial Fund, served on its board. The AICP had groomed John Kingsbury and Harry Hopkins early in their social work careers: Kingsbury, also Columbia-educated, had served with the American Red Cross in postwar France and just been appointed as the executive secretary of the Fund, and Harry Hopkins, then drafting the charter for the American Association of Social Workers and soon to be its president, would later return to New York City to direct the Fund's own demonstration projects before leaving to develop many signature New Deal programs.[23]

Under Burritt's leadership, the AICP also played a key role as a mediator between private philanthropy and public health. And it had experience doing so in health work. Before the war, the National League on Urban Conditions among Negros, the forerunner to the National Urban League, documented unprecedently high mortality rates among the residents of the Columbus Hill section of the city, a black tenement neighborhood between Amsterdam Avenue and West End Avenue, and 59th Street to 65th Street. The League, headquartered in New York City, championed the cause of black social welfare by black social workers.[24] But its expansive definition of social welfare also included the way some black nurses saw themselves. Adah Thoms, an 1895 graduate and then acting director of nursing at the black Lincoln Hospital Training School for Nurses, served as the League's secretary. Thoms received postgraduate training in public health nursing at Henry Street and in social work at the New York School of Social Work.[25] She was one of the most influential nurses in both New York City and the country. Edwin Embree, in charge of the nursing portfolio at the Rockefeller Foundation, recognized her training school as "one of the three first-rate negro training schools in the country," and Thoms as an "excellent chief."[26] The norms of segregation in New York City did allow her access to the best postgraduate education, albeit for work in her own community of color. But it excluded her from the inner circles of nursing leadership and barred her from ever holding more than an acting director title at a school that she, in fact, actively led.

In response, and in keeping with the city's commitment to health within a segregated system, Haven Emerson, then the city's commissioner of health, asked the AICP to begin health work in the Columbus Hill neighborhood. In 1916, Burritt brokered funding from a small, private philanthropy in Westchester County, New York, and set Clara Price, a black nurse, upon the task of reducing infant mortality rates of 155 babies per 1,000 to closer to the 35 per 1,000 among white babies in the neighborhood. By 1919, Burritt was reporting to his patron that the "fabulous return of successful industrial stocks cannot compete with the return on this investment." Of the 160 infants born the previous year,

only three infants died. A total of 208 of the 253 babies born in the previous eighteen months received physical examinations and nutrition support at
a new milk station; and those that did not had moved from the neighborhood.
The striking success of this work, Burritt concluded, led to the AICP decision to
desegregate its convalescent home for sick mothers and babies in Westchester
County. By the early 1920s Columbus Hill had increased its numbers to four
black nurses: Thoms had personally trained two of these nurses and all had
Henry Street postgraduate experience.[27]

Homer Folks joined Burritt in the interlocking circles of public health leadership. Folks, educated at Harvard and practicing as a social worker, had also
served with the American Red Cross (ARC) in France both during and after
the war. He returned to his position as secretary of the New York State's Charities Aid Association, charged with creating cooperative bonds between private
and public health agencies in all areas of the state except New York City. His
wide-ranging interests included mental hygiene, infant mortality, and, as the
presiding officer of the first White House Conference on the Care of Dependent
Children in 1909, child welfare. But his abiding interest was in tuberculosis
prevention and treatment. Folks was the first layperson elected president of the
National Tuberculosis Association and viewed with excitement the fact that
the Framingham Study's data might undo decades of his own work encouraging New York State's towns and counties to build tuberculosis hospitals.[28]
Visiting nurses in Baltimore, he told an AICP meeting in 1921, had presented
compelling data at the Sixth International Congress on Tuberculosis in 1908
that TB in one member easily spreads to other members of a family. He returned
from the Washington, DC, congress, he remembered, "weighed down by the
burden upon us, apparently of starting tuberculosis hospitals." Yet now, with
the Framingham data—and a commitment to "medical knowledge and nursing
supervision, and of actual aid and moral support"—one could keep families
together and enable their ill loved one to fight their second war with TB: that of
surviving, making a living, and keeping their households and affairs in order.[29]

Not surprisingly, Hermann Biggs, now commissioner of health for New
York State, joined Burritt and Folks at important policy tables. Biggs, trained at
Cornell University and Bellevue Hospital Medical School, had endured a bruising battle with the city's private medical practitioners in the closing decades
of the nineteenth century over mandatory TB reporting and monitoring. Biggs
had won that battle but lost the war. Concerned citizens, nurses, social workers, as well as private physicians were required to report suspected as well
as confirmed cases of tuberculosis to the city's Health Department. Each day,
the city's public health nurses were to receive the names of new cases in their

respective districts, visit them in their homes, and assess and report on the home's general cleanliness and sanitation. The nurse then taught families about proper sleeping arrangements, and the necessity of fresh air, good nutrition, cleanliness, nutritional access, the importance of not sharing eating and drinking utensils, and the proper disposal of expectorants. Ideally, the nurse would visit all patients without the means to pay private physicians monthly and, when necessary, she also had the power to begin a process for those poor families that could lead to the hospitalization in a sanitorium for an ill adult, or in a preventorium for an exposed child.

But, in reality, there were too few public health nurses, and those assigned to work with poor tubercular patients concentrated only on those deemed more difficult and recalcitrant. In addition, there were too many ways private physicians could evade reporting requirements for those with access to their care. And these physicians protested vehemently about how such requirements intruded on the confidentiality of the doctor-patient relationships. Biggs had learned to tread cautiously around the prerogatives of private medical practitioners even as what historians Amy Fairchild, Ronald Bayer, and James Colgrove call the "war time enthusiasm for controlling VD" led to congressionally backed monies for a new campaign for venereal disease detection and treatment in the 1920s.[30]

Livingston Farrand, the past chair of the ARC, had just returned to New York City from the University of Colorado to assume the presidency of Cornell University, as well as a seat on the board of directors advising Josephine Goldmark in her survey of US nursing education. But he was well known to leading public health reformers through his leadership positions in the National Association for the Study and Prevention of Tuberculosis and the American Public Health Association, serving as the editor of its *American Journal of Public Health* from 1912 to 1914. Farrand trained as a physician at Columbia's College of Physicians and Surgeons but pursued academic interests in psychology and anthropology as well as tuberculosis. He, too, served in postwar France for the Rockefeller Foundation's Commission for the Prevention of Tuberculosis in France. His associate director in France, James Alexander Miller, had been instrumental in establishing and then chairing the New York Tuberculosis Association when he returned. Miller had originally trained as a chemist, and eventually secured a position in the research laboratory of New York City's Department of Health. His work caught Hermann Biggs's attention. Biggs encouraged Miller to attend medical school at Columbia, promising to hold his position in the laboratory until graduation. Miller never returned to the laboratory; rather, he went into clinical practice and came to the planning of a new health center as well as a new campaign for tuberculosis control as one of the leading clinicians in the field of pulmonary diseases and tuberculosis.[31]

Planning a Health Center in East Harlem

The idea of health centers was not new to New York City. In the late nineteenth century, public health nurses, politicians, and philanthropists joined forces to create local infant welfare stations to promote clean milk and breastfeeding in the battle to decrease infant mortality.[32] In 1914, S. S. Goldwater, the city's health commissioner, opened a geographically defined health district in the Lower East Side to try to bring the work of the Health Department closer to the individuals in need. The success of this initiative led Goldwater's successor, Haven Emerson, to extend the concept to the entire borough of Queens, carving it into four new health districts. But, as was so often the case with early twentieth-century public health departments, a change in political administration brought a change in both health commissioners and public policy. The new city government leaders in 1918, abetted by health department bureau chiefs who saw their centralized authority diminished by local district administrators, abolished these health districts.[33] By the early 1920s, a once brilliant New York City Health Department had entered what its historian, John Duffy, describes as the "Years of Travail."[34] The *American Journal of Public Health* warned of a "disorganization" that would destroy the department's reputation, as political patronage rather than service delivery had become the function of the department.[35] Demonstration projects, like those being planned for New York City, would serve as bracing antidotes to this malaise.

Homer Folks, also a member of the Board of the New York County chapter of the American Red Cross, took the lead in turning its support of health centers into a reality. The national offices of the ARC had pledged a building, and, in keeping with the practices of other ARC-sponsored health demonstration projects across the country, the local chapter surveyed the various neighborhoods in the city. It eventually chose East Harlem, a defined geographic district recognized by the city's health department, as the site of the projects. It met criteria in that it was a defined local area of approximately 100,000 people with twenty-three private health and welfare agencies who agreed to cooperate by locating all their neighborhood offices in one centralized site. It did not meet criteria in that it hardly represented a cross-section of the city regarding health outcomes and standards of living.[36] Yet, the ARC chapter realized, in the socially, ethnically, and racially stratified world of New York City neighborhoods, there seemed no such geographically defined area that could.

The officially defined boundaries of East Harlem stretched from East 99th Street to First Avenue, East 104th Street to Third Avenue, Third Avenue to the Harlem River, and finally, from the Harlem and East Rivers to East 129th Street. It was home to what the city recognized as the "largest Italian colony in the western hemisphere."[37] It grew as late nineteenth-century immigrants from Southern Italy sought relief from the traditional but overcrowded ethnic neighborhoods of the Lower East Side. Yet, their standards of living hardly improved. Most tenements were dilapidated "old law" buildings: they had been constructed before newer building codes took effect and they had shared outdoor bathrooms and no running water. A few men worked as skilled artisans but most were employed as laborers, factory hands, or petty tradesmen; and one-third of its women had to supplement their families' incomes by home-work making paper flowers or by sewing factory-consigned garments.[38] With the postwar immigration restrictions of 1921 and 1924, their numbers were evenly split between those who were foreign-born and those born of foreign parents. And they and their babies died at rates greater than those for New York City as a whole. In the period between 1916 and 1920, adults in East Harlem suffered a 15.3 per 1,000 mortality rate as compared to 14.7 in the city; during the same time period, their babies died at rates of 100.6 per 1,000, rather than the 83.2 mortality rate for the city as a whole.[39]

East Harlem was also home to one of the most distrusted immigrant groups in the United States. Its inhabitants hailed from the poor, agricultural *Mezzo-giorno*, the southernmost part of Italy that was connected, in the public imagination, with notions of pervasive superstition, illiteracy, dishonesty, violence, and crime.[40] These individuals' darker complexions; deep suspicions of institutional authorities that had only oppressed them in their native land; emotionality; particular dialects and religious practices; devotion to the *domus* (or family) above nation; and ritualized, hierarchical patriarchal practices that left little room for self-expression raised profound anxieties about race, assimilation, norms of citizenship, and proper gendered relationships.[41]

Folks and his colleagues dreamed of control: of finally having an opportunity to rationalize a "criss-crossy" and inefficient system of private philanthropy and public health that brought material resources and healthcare to the homes of the deserving, and frequently tubercular poor.[42] But they knew that had to start incrementally. They had to first work on a system of care coordination: of bringing all the East Harlem neighborhood's health and welfare agencies together in one building for "one-stop shopping." In their minds, each agency would maintain its own budget and administrative structure. But they hoped to "demonstrate" that there would be increased service utilization and

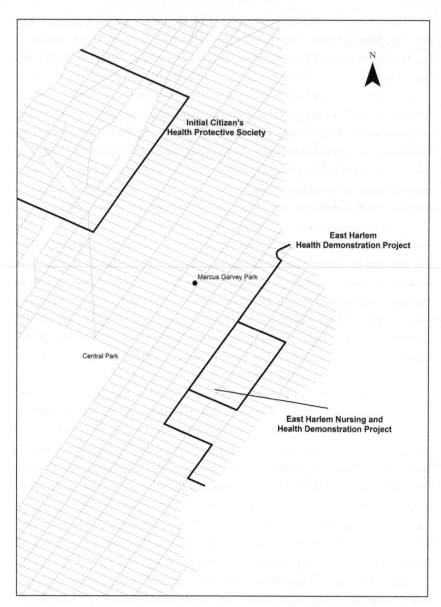

N

Initial Citizen's
Health Protective Society

East Harlem
Health Demonstration Project

Marcus Garvey Park

Central Park

East Harlem Nursing and
Health Demonstration Project

Figure 1. Map of the East Harlem Health and the East Harlem Nursing and Health
Demonstration Projects Neighborhoods

Source: Center for Population Economics (2013), The Historical Urban Ecological GIS Data portal—New York City, HUE Manhattan Street Centerlines c. 1930 [Data set], http://ue.uadata.org/gis/. Retrieved October 2, 2015. Map created by Lisa Hilmi.

better health and welfare outcomes when services were more physically accessible to those in need.

But increasingly, the social reform leaders of almost all private organizations had nightmares about six or seven nurses and social workers descending on one unfortunate family—each interested in a part of a problem rather than one worker approaching the family as a whole.[43] A poor family that included a father with tuberculosis, a child with pneumonia, and a new infant, for example, might be visited by a city health department nurse with experience in tuberculosis, a VNS nurse to provide the bedside nursing of the sick child, an MCA nurse to supervise the health of the new mother and baby, and an AICP social worker to help the family obtain good nutrition and material resources.

Thus, in addition to demonstrating the value of health and social welfare care coordination through a health center, New York City's unique contribution to the national health center movement would be a subsidiary demonstration involving nurses that would be one of care control. All of the city agencies providing nursing services in the neighborhood of East Harlem—the Department of Health, Henry Street, MCA, AICP, and Saint Timothy's League of lay women supporters of public health nurses—would pool their resources, personnel, and dollars into one controlling organization with its own governing board.[44] The intent would be to prove that a unified nursing service could deliver more effective, more efficient, and more intensive care to more families than the prevailing fragmented one. But there was a problem. Both the physician S. Josephine Baker, then head of the city's Bureau of Child Hygiene that employed the predominant number of nurses in the Department of Health, and Lillian Wald said no. They immediately dismissed the idea of anyone else supervising their nurses.

Folks knew that bringing the city's own public health nurses into a centralized service would be a long process. Unlike nurses in other private agencies who had to rely on their powers of persuasion, the city's nurses had official police authority to identify and report suspected cases of communicable diseases, to institute quarantines, inspect homes, and provide continued surveillance of those receiving treatment in the city's clinics. Wald's resistance was not unexpected but, unless managed well, threatened to derail their grand plans. The officials at the Rockefeller Foundation bluntly told them that the participation of Henry Street, which it also financially supported, was "key" to any Foundation support of the demonstration project. They also offered a strategy. "Confidentially," Foundation officers suggested, it would be Annie Goodrich, now the current director of the VNS at Henry Street, rather than Wald, who they reported as retired, who needed to be "won over."[45]

Wald had withdrawn from the day-to-day operations of Henry Street, but she had hardly retired. She immediately reengaged, expressing objections to two aspects of the nursing demonstration. She believed that the highly respected work of her Henry Street nurses would be compromised if outside nurses were to supervise their practices: that they would not be held to the same high standards that governed Henry Street's bedside nursing care. More significantly, she objected to plans to use the nursing demonstration to research one of the most hotly debated questions in national public health nursing practice. Should the organization of practice be built on a "generalized" model where one nurse met all the nursing and health needs of a defined neighborhood, as it was at Henry Street? Or should it be organized around a "specialization" model where one nurse developed the knowledge, skills, and techniques in defined areas such as tuberculosis, maternal-child, and school nursing—as it was in the city's own health department? Wald declared that question had already been definitively answered: Generalized nursing practice had definitively proven its worth. To investigate further wasted time and resources.[46]

Folks called in reinforcements. He reached out to Burritt and John Gebhart, the AICP's director of social welfare. Burritt sincerely believed in generalized nursing as "the future thing."[47] But he also knew of the significant debate that still existed in public health and public health nursing about how its practices should be organized. A case for specialization in public health nursing could still be strongly argued: It was simpler in its organization; its practitioners were more easily supervised; and it did not lose the important work of health education to the priority demands for nursing acutely ill patients at home. Specialized public health nurses, particularly in areas such as tuberculosis nursing, also had more expert knowledge and more skilled techniques that contained cross-infections.[48] Moreover, he was stunned that Wald claimed "complete surprise" at the care control aspect of the proposed nursing demonstration. The Henry Street Board had approved the proposed demonstration, but Wald and Goodrich stated they believed that they had only pledged participation. Burritt and Gebhart had a lengthy conversation with Wald and Goodrich that involved both "Homerian diplomacy" and the bluff that if Wald would not agree to let Henry Street join the demonstration, "it would go forward without her cooperation and that she would, therefore, be left in the position of opposing an important future development in the nursing field."[49] Wald and Goodrich finally agreed to participation, but it came at a price. Maternity Center Association nurses had to withdraw their prenatal work from East Harlem homes, leaving all home visiting within the domain of Henry Street nurses in East Harlem.[50] With a building donated by the American Red Cross and funding promised by the Laura Spelman Rockefeller Memorial, a trust of the Rockefeller Foundation dedicated to supporting the health and welfare charities of the late wife of John. D. Rockefeller Sr., the planned demonstrations could become realities.

Planning a Health Demonstration in Bellevue-Yorkville

Burritt again faced resistance from Lillian Wald when he approached her about Henry Street nurses' participation in the Milbank Memorial Fund's planned health demonstration in the Bellevue-Yorkville section of the city. Burritt has secured financing from the fund to resume a small home hospital in an available apartment building in 1919, and by 1921 he dreamed, along with John Kingsbury, his protégé and the new executive secretary of the fund, about plans "to home hospitalize entire sections of the city."[51] The Framingham Study was, in Kingsbury's opinion, one of only "average" dimensions. What he was seeking in the home hospital concept of how to care for families when a member had tuberculosis was a "monumental enterprise" for the $2 million the Fund stood ready to invest in health, particularly public health, work.[52] Public health nurses would yet again be critical to the success of this kind of demonstration. But now they would incorporate specialized tuberculosis nursing practices into their generalized work—they would bring into their practice one form of communicable disease care that had long lay outside it.[53] This was a plan long advocated by leading public health nurses across the country. Elizabeth Fox, the national director of the American Red Cross's nursing service and an invited participant at the 1922 Milbank Fund's Advisory Council Annual Conference—a council of national experts that advised the fund about the direction of its planned demonstrations—supported this plan. "I saw in your program somewhere the necessity of adding tuberculosis nurses," she reported, "and it immediately struck my eye as one of the things I hoped you were not going to do."[54] But the issue for Wald was, again, not practice but her own control. Five months after her initial discussions with Burritt, she finally agreed to let her Henry Street nurses participate in the Bellevue-Yorkville demonstration if they had their own Henry Street supervisor.[55]

But in this instance, Wald was the least of Burritt and Kingsbury's problems. At a March 1923 conference convened by Albert G. Milbank, the president of the Fund, to consider a plan for the control of tuberculosis in New York, Hermann Biggs, now commissioner of health for New York State, laid out all his objections to Burritt and Kingsbury's proposed enterprise. If he had extra money, would he spend it on TB work? No he would not: TB deaths, along with those for typhoid fever, infant mortality, and diphtheria, had been steadily declining in the state, but deaths from cancer, kidney, and cardiac diseases had been as steadily increasing. "It is in the degenerative diseases of middle and advanced life," he told his audience, "that the advances are to be made in the future." A tuberculosis campaign had value, he explained, but only as a "text" that the wider constituents of the general public and private medical practitioners understood and supported. The real lesson of the Framingham Study, he concluded, involved the importance of a general medical exam for all men, women, and children that identified and treated presumptive and actual cases not only of tuberculosis but also of those new diseases responsible for increasing mortality rates. Biggs had been trying to promote the value of general medical exams in New York City and State for years, but with minimal results. Doctors of the "old day" would not accept this new practice; and their potential patients remained suspicious that this was but another initiative to line physicians' pockets with extra money. In Biggs's considerable and influential opinion, a demonstration that would influence patients to demand and private medical practitioners to provide general medical examinations would be of "inestimable value." More concretely, he predicted, it would also add ten to fifteen years to the average American's life expectancy.[56] By 1924, any reference to tuberculosis had been dropped from plans for the demonstration project planned for the Bellevue-Yorkville section of New York City.[57]

The Bellevue-Yorkville demonstration was itself nested within a series of three demonstration projects the Fund supported in New York. It had appointed a Technical Committee that included Biggs, Burritt, Folks, Farrand, Miller, and Haven Emerson, who also sat on the board of the East Harlem project. The committee's charge was to operationalize the Fund's commitment to "demonstrate" in three different—yet typical for their size—communities that the intensive use of "all known health measures" would substantively reduce rates of mortality and morbidity at a cost the community would willingly pay to ensure the continuation of the demonstration's initiatives when Milbank funding ended.[58] It was also charged with treading lightly around the concerns of private medical practitioners. As Edward Baldwin, a Massachusetts physician involved with the Framingham Study, recounted to the committee, private practitioners had no problem with tuberculosis demonstrations: these physicians rarely treated poor tubercular patients, who were instead sent to sanitariums. But any hints that a demonstration might attack rates of typhoid fever, scarlet fever, and other diseases would bring howls that an outsider "will take the bread out of my mouth."[59]

In 1922 the committee chose Cattaraugus County, in rural upstate New York, and Syracuse, a mid-sized city also upstate between Albany and Buffalo, as the sites of its first two demonstrations. The third, in a "metropolitan area,"

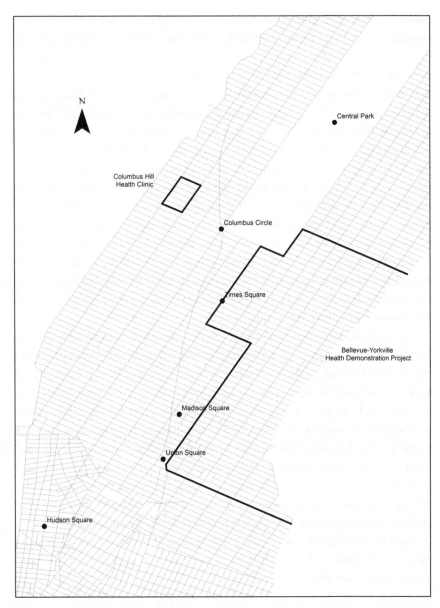

Figure 2. Map of the Bellevue-Yorkville Health Demonstration Project Neighborhoods

Source: Center for Population Economics (2013), The Historical Urban Ecological GIS Data portal—New York City, HUE Manhattan Street Centerlines c. 1930 [Data set], http://ue.uadata.org/gis/. Retrieved October 2, 2015. Map created by Lisa Hilmi.

presented more problems. Brooklyn wanted to host the demonstration, but the prestige lay in a site in Manhattan. By 1924, the committee had decided on the Bellevue-Yorkville sections of the city. This was a two-square-mile area in the central east side of Manhattan between 14th and 64th Streets and from the East River to 4th Avenue for most of the district, but slipping up to 6th Avenue in its northern section to include established tuberculosis clinics in Yorkville as well as Bellevue. In some ways this was an odd choice: Census data about mortality and morbidity were deemed unreliable as the district's populations had been sharply declining in the face of rapid commercialization. In addition, the district was flooded with workers during the day who returned home at night to other areas of the city or suburbs. But its strength was seen in the economic diversity of its population ranging from poor and working-class families in Bellevue to middle-class and even some upper-class families living in new apartments in Yorkville. Most importantly, however, the Bellevue-Yorkville district was rich in hospitals, including some of the most respected ones in the city, outpatient clinics, and private medical practitioners—the key constituents the demonstration wanted most to reach.[60]

Yet, it faced immediate resistance from an unconsidered constituency: the people it sought to serve. The Milbank Memorial Fund's 1924 announcement that it would spend $2 million to improve the health of the citizens of New York was met with some skepticism in an article in the *New York Times*. Will they want to be helped, it wondered. Are they willing to "be lessoned in these things by strangers from the outside"? The implication of the announcement, it continued, "is that at present the habits of these people are bad or at the least unwise in matters of sanitation and hygiene, but it has not been reported that they have made any such admission or appealed for the reformers to come among them and improve their condition." There was only one thing that merited this intrusion of privacy, it concluded. In the tenements, "most of the mothers have learned that a visiting nurse, though unmarried, may know about the care and feeding of babies a thing or two that was not known by mothers and grandmothers in 'the old country.'"[61]

Chapter 2

The Houses That Health Built

On November 10, 1921, New York City's East Harlem Health Center Demonstration Project (Health Center) opened to great fanfare. The Health Center self-consciously characterized itself as a "department store of health and welfare" playing on the success of a new middle-class institution that promised everything one could imagine buying in one central location. Similarly, the Health Center gathered twenty-three of the neighborhood's health and social welfare agencies into one newly refurbished building for the same kind of "one-stop shopping" for coordinated health and welfare services. The concept of "coordination" was key to the success of the Health Center. Public and private agencies would keep control of their budgets and personnel; but the demonstration would test the premise that physical proximity would eliminate costly service duplication, ease access to resources needed by the predominantly Italian community, and, in the end, deliver better health outcomes.[1]

A little more than a year later, in December 1922, its sister demonstration, the East Harlem Nursing and Health Demonstration Project (Nursing Project), started its work with less attention but no less import. Unlike the Health Center, the Nursing Project would be an effort in controlling the distribution of nursing services in one-half of the East Harlem neighborhood. The three private agencies that supported specialized East Harlem nursing services—the Henry Street Visiting Nurse Service (VNS) that focused on nursing the sick in their homes, the Maternity Center Association (MCA) that provided prenatal and home birth services, and the Association for Improving the Condition of the Poor (AICP) that supported tuberculosis nurses—would pool their resources, personnel, and dollars into one controlling organization that would also construct

a research project to prove that a generalized nursing service could more efficiently and effectively serve the needs of the neighborhood for sick nursing,
provide maternal healthcare and education, and meet the health needs of the
preschool child.[2]

And, finally, in 1926, the Bellevue-Yorkville Health Demonstration opened
in midtown New York City. It had—with some "apprehension"—refocused its
goals and agreed to an administrative arrangement that placed the health commissioner in charge and a member of his staff as the director of the demonstration. Within one year of its opening in 1927, however, the Fund found this
arrangement "impossible," with vague allusions to the "handicaps" of working
within the structure of the city's "political machine." It again reconfigured its
mission as a smaller series of demonstrations, some of which—like the use
of chest X-rays in the diagnosis of tuberculosis and the provision of materials needed to maintain lung rest through induced pneumothorax—could be
adopted later by the Health Department.[3]

This chapter delves more deeply into the day-to-day realities of New
York City's health demonstration projects. It explores the escalating tensions
between New York City's Department of Health and private agencies and associations over who controlled the public health agenda. These private or, as they
referred to themselves, voluntary agencies publicly ceded control to the official
agency that the Departments of Health represented. But privately they constantly sought ways to turn this official agency toward their priorities. In New
York City, both the Rockefeller Foundation and the Milbank Memorial Fund
believed public health nurses were key to this process. Indeed, the involvement
of the city's public health nurses in both East Harlem demonstration projects
had been a central element in the Rockefeller Foundation's support. It could
not be a true demonstration of care control, the Foundation believed, unless
it involved the city's own public health nurses who ran the milk and infant
welfare stations; who supervised the health of schoolchildren; and who implemented programs of case finding, case holding, and case control of tuberculosis
and other infectious diseases. The Foundation's policy, in the United States and
abroad, was one of only working through governmental public health authorities to ensure the sustainability of its initiatives. It hoped to use a consolidated
private and public health nursing system in East Harlem to ultimately do the
same in New York City.

Historians have long noted the tensions between public and private agencies in setting and implementing a public health agenda.[4] But public health
nurses held no interest in the battles at tables to which they had not been invited.
More precisely, the nurses involved in New York City's health demonstrations

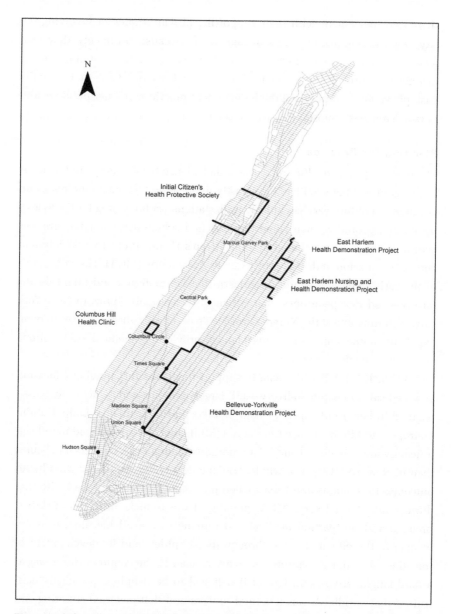

Figure 3. Locations of All of Manhattan's Health Demonstration Projects and Clinics circa 1925

Source: Center for Population Economics (2013), The Historical Urban Ecological GIS Data portal—New York City, HUE Manhattan Street Centerlines c. 1930 [Data set], http://ue.uadata.org/gis/. Retrieved October 2, 2015. Map created by Lisa Hilmi.

shared no investment with their supporting philanthropies in involving the city's own public health nurses in their work. Because, ultimately, they won what they themselves wanted. By the end of the formal demonstration period in 1928, both private and public health nurses in New York City—not, as in the past, physicians—supervised the independent practices of other public health nurses. This was a substantive achievement.

Planning for Practice

The postwar public health practice agenda had also turned its eye to the needs of two groups it believed had been vastly underserved: pregnant mothers, and children too old for services at baby milk stations yet too young for the assessments they would receive when they reached school age. Prenatal care was central to the services offered by the Citizen's Health Protective Society and core to the mission of the Maternity Center Association (MCA). The MCA, born of the early twentieth century's concern that US maternal and infant deaths far exceeded European ones, had grown to thirty small clinics in New York City, including one at the Nursing Project, These clinics offered classes to poor, expecting mothers and hired physicians to provide the medical examinations and treatments they needed.[5]

By 1921, MCA had the data to support its claims that nursing education and medical care before delivery—still largely at home and by lay midwives—resulted in better outcomes for both mothers and their infants. Louis Dublin, Metropolitan Life Insurance Company's (MLI) chief statistician, had found significantly lower maternal and infant mortality among MCA mothers and babies than in New York City as a whole. Looking deeper, he also found even better outcomes for mothers and babies when prenatal care was combined with that during and after delivery.[6] MCA's grand goal was to make this kind of skilled prenatal and postpartum medical and nursing care available, first, to every woman in the city, and, later, through its pamphlets and its development of traveling educational institutes, to every woman in the country.[7] But it waged a hard-fought drive against what it believed to be deeply entrenched "half-truths" that childbirth was a natural process and that maternal suffering and death were but God's will. When will mothers and fathers realize, MCA's officers wondered, that the entire pregnancy experience through and after delivery subjected the mother "to such a strain that the margin between health and disease becomes dangerously narrow and the balance can only be obtained by constant supervision and care." MCA, however, stopped short of completely medicalizing childbirth. It envisioned, and later implemented, a new public health worker. It had begun to lay the groundwork for a "nurse-midwife," a skilled trained nurse with postgraduate education in obstetrics that would replace the traditional midwives upon whom poor mothers depended.[8]

At the same time, rising concerns about the physical "defects" found in young American men during draft examinations, and a seeming epidemic of malnutrition among schoolchildren in the immediate postwar period, focused attention on what Yale professor and pediatrician Arnold Gesell had characterized as a vast "wasteland" in public health practice: the health of the preschool child.[9] Too often, it now seemed, it was only a schoolchild's first health exam that discovered the rampant degree to which children suffered from such "defects" as infected tonsils and adenoids, carious teeth, and poor vision. Physicians, as historian Jeffrey Brosco has argued, believed such defects caused childhood malnutrition. And they felt that some cases of childhood malnutrition laid the groundwork for diseases now on the public health agenda radar: tuberculosis, congenital heart disease, and syphilis.[10] But in the absence of any support for universal nursery schools as was implemented in England and increasingly prominent on the Continent, the problem lay in finding these children at home, bringing them in for treatment, and then teaching their mothers about the importance of teeth brushing, good nutrition, outdoor activities, and developmentally appropriate play. And the solution, yet again, was public health nurses who visited such families on their daily rounds.[11]

Running the Demonstrations

In addition to the existing health and welfare services, the East Harlem Health Center's three-year demonstration period saw a dizzying array of health and welfare services coordinated, created, and consolidated. Under the leadership of Kenneth Widdemer, the center's executive director, and with the input of a House Council of representatives of the community and affiliated agencies that ensured its "democratic character," the Health Center canvassed the neighborhood families to learn about their perceived health and welfare needs and to make them aware of the resources available at the Health Center. It also strengthened its focus on getting adults access to general medical exams as a way to identify potential problems before they became serious diseases. It worked together to establish new cardiac clinics to address what was quickly becoming one of the leading causes of adult mortality. The Health Center also collaborated with the city to provide physical examinations and dental services to East Harlem's schoolchildren about to enter school; reorganized TB prevention work to more closely follow children deemed at risk; and systematized record keeping.[12]

The services offered, Widdemer admitted, were by no means comprehensive. To accomplish a "complete health job" would have been so prohibitively

expensive that the city's Health Department would refuse to assume any responsibility for its continued existence after the demonstration ended. Moreover, he
continued, the true purpose of the Health Center was as a demonstration that care
coordination improved health outcomes.[13] Indeed, one of the key features that
attracted the attention and the financial support of both the Rockefeller Foundation and the Milbank Memorial Fund was the health demonstration projects'
commitment to carefully documented metrics about its community's health and
welfare. One of the most critical social issues in the decades after the First World
War was the rapidly escalating costs of medical and nursing care.[14] And the demonstration projects' potential to generate data to improve access, decrease costs,
and develop models of effective care that could be used across populations won
important philanthropic support that bolstered additional funding from private
agencies and public health departments across the country.

The initial three-year data from the Health Center seemed impressive. By
1924, it had served 33,000 individuals in a neighborhood of some 112,000. The
calculated costs per capita rose somewhat above the city average to twenty-
seven cents. But the numbers of individuals served rose 109 percent. While
rates of tuberculosis remained much higher in East Harlem than in New York
City, number of deaths from this disease was approaching that of the city as a
whole: 38 per 100,000 died from tuberculosis in East Harlem versus 37 in New
York City. Infant mortality declined 36 percent by 1923 as compared to the
city's decline of 25 percent. And the death rates from all causes of mortality
now mirrored that of the city at large: 1,176 per 100,000 in East Harlem and
1,171 per 100,000 in New York City.[15]

The Nursing Project, under the direction of Grace Anderson, formerly head
of the Municipal Nursing Service in St. Louis, flourished as well. It launched
a well-designed comparative study of the effectiveness of generalized versus
specialized nursing with carefully matched East Harlem neighborhoods organized either to receive care from an array of nurses specializing in maternity,
infant welfare, preschool, and sickness care or else to receive care from one
nurse responsible for the health needs of an entire neighborhood. Anderson
had started the Nursing Project with a nutritionist responsible for working
directly with families whose children were identified as malnourished. But she
had quickly switched to a system in which the Project's nutritionist served as
a consultant to the neighborhood nurses who would now incorporate nutrition
work into their generalized practice. Anderson reworked the Project's recordkeeping system to also include length of time of nursing visits to enable another
study that would compare the costs of different kinds of nursing home visits.
And, she had plans in place to study the nutritional status of children with

pneumonia, a group who made high demands on its bedside nursing service. Anderson's and her public health nurses' work, often invisible in the published reports, substantively contributed to the district's impressive outcomes. And their faces were those most often seen in the neighborhood: two years of data documented 63,500 visits to individuals and families throughout the district.[16] In 1925, the Rockefeller Foundation approved funding for two more years of both demonstration projects.[17]

The Nursing Project was a bright moment for New York City's public health nursing leaders. In 1924, they had to close the doors of the Citizen's Health Protective Society. The tensions between nurses from the Maternity Care Association and the Henry Street Visiting Nurse Society over who was better equipped to provide prenatal care were never fully resolved. A tentative agreement had MCA nurses providing all prenatal work and continuing the postpartum care of members who delivered in hospitals, while Henry Street nurses would provide a nurse at the time of delivery for other mothers who delivered at home; Henry Street nurses would continue care through their mothers' postpartum period. But this agreement foundered on the Henry Street director's wish to exert leadership. If, Annie Goodrich tersely informed Olive Husk, the Society's nursing director, a Henry Street nurse met a pregnant mother when in the home delivering sickness care, Henry Street would continue with the family doing the needed prenatal work.[18]

But, in the eyes of the nurses, the most serious problem involved the Manhattanville families themselves. Mothers wondered why they should pay for prenatal services they could access for no cost at a nearby Department of Health clinic. They also quickly realized that they could join the Society right before an expected delivery and pay a membership fee that was less than what they would have to pay a Henry Street nurse for care during and after their delivery. They learned to take advantage of a new installment membership fee structure: They would pay one-quarter of a family membership when someone fell ill and then never continued to pay the rest of their membership dues when wellness returned. And, as one young mother frankly questioned, why should she pay for something she did not need? If she were to fall ill she would not need a nurse for a few hours per day; she would need someone to look after her very young children.[19] Nurses prioritized health services; mothers also wanted housekeeping ones.

As alarmingly, the Manhattanville neighborhood was itself changing. As Husk wrote Goodrich in 1922, there had been an increase in the number of inquiries from black families about the services the Society offered. While Husk consistently "discouraged" such inquiries, there still existed a distinct "danger" that what was to be a white, middle-class insurance program would change into one with a "larger colored service" because of the increasing presence of black families in the neighborhood. Husk and Goodrich shared the assumption that a segregated Society would be the only way to attract the white middle-class families they sought. The Society moved farther uptown in late 1922 to what seemed to be a more promising location at 134 Street and Amsterdam Avenue. This community remained uninterested as well. Husk and Goodrich continued to blame families for the Society's failure. "Perhaps," the nurses wondered, "in attempting to popularize a new development, we selected a most difficult district where community spirit and pride and cooperation are little thought of."[20] Yet, the reality was that much colder. While families appreciated health work, they would only pay for illness care. They would not pay for nursing healthcare.

And fissures were emerging in the Health Center's plans for cooperation. In 1924, the American Red Cross abruptly announced its withdrawal from the national health center movement and now assumed that local agencies would take on increasing financial responsibilities and administrative costs.[21] And the agencies themselves were rethinking their commitment to coordination. Some had had to redraw their own long-established practice boundaries in New York City to conform to those of the Health Center and others found themselves providing more resources to the families in East Harlem than they did for those in other neighborhoods of the city. Certainly, as Homer Folks explained to Beardsley Ruml, the new director of the Laura Spelman Rockefeller Memorial, in May 1925, there had been some "misgivings" on the part of some participating organizations when plans were first presented to them. But, he continued, the ultimate success "was even greater than anticipated" and all had agreed to continued participation past their initial three-year commitment.[22]

But Ruml had begun to hear otherwise. Ruml moved into the tight circle of early twentieth-century philanthropists at an early age. At twenty, he served as the assistant to the president of the Carnegie Foundation; a few years later he served as an advisor to the Rockefeller Foundation; and, in 1922, at the age of twenty-seven, he was appointed to create a more focused philanthropic vision for the Foundation's Memorial. At his urging, the Memorial had already begun to move away from its tradition of funding health and social welfare projects— moving away from funding individual private agencies such as Henry Street, the Maternity Care Association, and the AICP—and toward a more sustained program of grant support for initiatives in the social sciences.[23] He knew of the difficulties facing the Milbank Memorial Fund as it tried to launch its own New York City demonstration. The "monumental enterprise" had been to build on the success of the AICP's prewar "home hospital" demonstration. But this depended on the cooperation of the city's Department of Health—which refused to subsidize the treatment of any adult at home as long as there were empty beds in the city's TB sanitoriums; the commissioner of health—who made it clear that any such initiative had to be "subordinate to" his authority; and to participating health and social welfare agencies—who were quickly losing interest given their experiences with care coordination in East Harlem.[24]

Ruml launched his own survey on the state of the Health Center in early 1926. If there were, as internal memos noted, a "spirit of cooperation" among those actually working at the Health Center, this did not hold true when discussing the center with the leadership of the participating organizations.[25] Issues of privilege and prerogative, colored by class and religious biases, undermined prospects of real cooperation. Certainly, Lawson Purdy, the director of the city's Charity Organizing Society (COS), another of the city's private social welfare agencies, had deep reservations; it actually cost more, he explained, to keep his organization with the Health Center because his social workers were "of higher intelligence and better trained" than those from other organizations and they wasted a great deal of time correcting the mistakes of other agencies' workers.[26] Lillian Wald, speaking confidentially, felt the Health Center accomplished little, was very badly organized, and, as it charged nothing for its services, pauperized patients. And, she noted, she thought as little of the Milbank Memorial's project in Bellevue-Yorkville.[27] In addition, as Folks did admit, Catholic relief organizations contributed little to the Health Center, placing spiritual values above social welfare work; they were also, he reported, quite content to have the secular AICP take on their cases.[28] And Burritt, when carefully questioned about how the health statistics differentiated the work of nurses from that of other workers in the Health Center, found he could not answer. The numbers, he conceded, were "all jumbled together."[29]

The case against the Health Center continued to mount. In 1926, Louis I. Harris, the city health commissioner, announced plans to form a new Welfare Council of New York City, an organization that would eventually bring together 332 of the city's largest health and welfare agencies for advice and consultation.[30] The medical and policy advisors to Ruml strongly recommended abandoning East Harlem and supporting the initiatives of the Council as they emerged.[31] In all likelihood, Ruml needed little encouragement. In a tactful letter to Homer Folks in April 1927, he explained how the Memorial had stopped funding projects in public health and public health nursing but that it was aware of the Memorial's historical commitment to New York City.[32] The Health Center received bridge funding to mitigate the impact of its closing on the East Harlem community until 1931 when the city took possession of a Health Center that had devolved into a lay-run Health Shop that dispensed health education pamphlets and created window displays.[33]

The Nursing Project, however, hoped to continue and, freed from "jumbled up" measures of its work, begin its journey toward what it would later call "a new approach" to health work.[34] In early 1927, aware that dedicated funding from the Memorial would stop in December, the leadership of the Nursing Project convened a "Continuation Committee" of its most important constituents. Chaired by Bailey Burritt, it included Hazel Corbin, the director of the Maternity Center Association and a leading voice in the campaign to train nurses as midwives; Florence Johnson, the director of nursing service of New York City County's American Red Cross; Margaret Nourse, the president of Saint Timothy's League, a group of laywomen supporting the Project; Marguerite Wales, now the director of nursing at the Henry Street VNS; Alta Dines, the AICP's nursing director; and Folks, in his capacity as secretary of the New York State Charities Aid Association and liaison to the Memorial. It also included Grace Anderson, and her assistant, Mabelle Welch, from the Nursing Project. A new constituent, Lillian Hudson, an assistant professor of nursing education at Teachers College (TC) at Columbia University, also joined the group. Students, including those from TC, were an increasing presence in the Nursing Project.

The Committee reviewed the Nursing Project's impressive accomplishments as it prepared to construct an argument about why it should continue even if the Health Center would not. Its research projects had produced data that contributed to the ongoing debates in public health nursing practice. Its published data supported generalized nursing as the best model for public health nursing practice. In this particular study, generalized nursing practice had outcomes as good as more specialized practices and generalized nursing was more efficient and cost-effective.[35] And it had worked with other leading public health nursing agencies to push the boundaries of what kinds of diseases and illnesses would be incorporated into generalized nursing practice. Like nurses at the Henry Street Settlement, East Harlem's nurses now cared for malnourished children, individuals with tuberculosis, and, in striking contrast with their earlier twentieth-century predecessors, those with communicable diseases.

These initiatives did not go unnoticed by other public health disciplines. In New York City, Lucy Gillett, the AICP's lead nutritionist, felt the move by the Nursing Project to incorporate nutrition counseling into its nurses' work "has hindered work in nutrition." As she wrote Burritt in 1935 summing up a decade of observations, the nurses' teaching was "perfunctory," inadequate in difficult cases, and served as a "bad model" for other agencies. She believed the prevailing sentiment among public health nurses was: "If East Harlem can do it so can we."[36] And the generalized practice of a public health nurse moving, for example, from nursing a child with the measles to teaching a new mother how to care for her infant raised such pressing questions about the potential to spread infections that a special forum at the 1925 Annual Meeting of the American Public Health Association had to be convened. What statistical evidence do we have of such cross-contaminations, the forum queried? None, Wales and other nurses and physicians responded. The real question that emerged from the forum was not one of statistics but of technique, in general, and bag technique, in particular. Both involved scrupulous hand washing with soap and fresh towels; technique extended to the uniforms nurses wore and the extent to which they physically interacted with others in the family; bag technique meant protecting the visiting nurses' bag with newspapers and always using freshly washed hands to retrieve objects within it. According to Wales and Dines, all techniques were carefully taught and practiced in New York City.[37]

The Nursing Project's other research involved the costs of different kinds of public health nursing care. Such data were essential to agencies that had to project budgets, if public, and determine fund-raising drives, if private. It reworked its recordkeeping system to also include length of time of nursing visits to enable another study that would compare the costs of different kinds of nursing home visits. Throughout 1924, its nurses kept detailed records of who they visited, for what reasons, how much time they spent in the home, and how much time they spent on other tasks such as travel, clinic work, and record keeping. Not surprisingly, the data found that postpartum care—care which also involved that of newborns—cost the most per visit ($2.96) because of the length of time involved (forty-six minutes per visit); sickness care followed, costing $1.62 for twenty-five-minute visits.[38] Surprisingly, the cost of teaching public health nursing students was not recouped by the services they rendered.[39] Students were expensive.

As the Committee reviewed the Project's accomplishments, it became clear that a teaching mission had slowly grown up alongside its service one. Over the past years, it had hosted increasing numbers of public health nurses from around the country; international nurse fellows supported by the Rockefeller Foundation; and postgraduate public health nursing students from TC. As it looked to the future, the Project envisioned expanding its service mission and formalizing its teaching one.

The Committee reveled in the Project's excellent service reputation. Homer Folks, when reporting to Ruml about the status of the Project, repeatedly emphasized how its nurses broke through an easy sense of futility when assessing the almost overwhelming health needs of the people of East Harlem with a series of "experimental programs" that magnified its impact.[40] As early as 1925, May Ayres Burgess, commissioned by the Foundation to do a qualitative study of specialized and generalized forms of nursing in the Project, found the work in both models of public health nursing practice to be "of a high grade," with "uniformity and poise in the excellent technique," and "thoughtful and intelligent" in their teaching. She found those nurses who worked under the specialty model more informed with the facts; yet those in the generalized model seemed more informed on the families themselves. Still, she wondered, was it possible that the Nursing Project "overemphasized" technique to the point where families could not follow the nurses' example?[41] This was an important question as families assumed increasing responsibilities for containing the spread of infections within their own homes.

The Nursing Project also had the strong support of Mary Beard, a powerful presence in public health nursing and now the assistant director of the Rockefeller Foundation's Division of Nursing Education. "It seems to me," she noted in 1927 after visiting the Project with the president of the American University in Beirut, it was "far and away the best place to observe health work for mothers and babies in New York. . . . One might easily have spent a week going from home to home with a public health nurse and not have seen so great a variety of health instruction as we saw that morning."[42] The Nursing Project's global footprint was an important strength. As Burritt wrote to John Kingsbury in 1928, Synneve Eikum, the US consultant to Brazil's first health center in São Paulo, "puffs with pride" every time her center introduces a change that can be traced back to the Nursing Project in East Harlem. Burritt felt confident in describing the Project as "the world's model health demonstration."[43]

By 1927, the Project had answered all calls for bedside nursing; reached 30 percent of all expectant mothers through a new infant service; and had 40 percent of preschool children under its health "supervision."[44] It saw a fivefold increase in costs during the past five years, but the families served increased fifteenfold and, despite its commitment to bedside nursing, it maintained a heavy financial investment in their health work with mothers and preschool children.[45] It had also established its place in the wider nursing community. Grace Anderson published an article describing the Project in *Public Health Nursing* in 1923; and one of the Project's clerical workers invited readers of the *American Journal of Nursing* to understand a day-in-the-life of a dedicated nursing service.[46] As importantly, it published pamphlets for widespread distribution to other public health nursing agencies on such topics as the lesson plans it developed for its clinic classes and the procedures it used to incorporate tuberculosis nursing, nutrition work, and services to preschool children into its program of generalized nursing practice.[47]

The Nursing Project had solved smaller problems as well. It had navigated tensions among its nurses themselves as different nurses from different organizations had different salary structures, different daily time schedules and vacation allotments, and, as importantly, different public health nursing uniforms.[48] It continued to negotiate tensions with Henry Street as the "vexed question" of charging fees to patients was continuously raised by the VNS.[49] It took to heart Burgess's one critique of its recordkeeping system: that it tabulated a visit to one family, for example, with three members under its care as three visits to individuals rather than as one visit to a family. This had important implications when calculating costs per visit, as care to three members of one family mitigated traveling expenses. Internally, the Project began cross-indexing families with individuals in an increasingly elaborate recordkeeping system; externally, it continued to report on individual visits as was normative in public health nursing practice.[50]

As it looked to the future, the Continuation Committee hoped to expand the Nursing Project to the entire East Harlem district and to add more work for preschool children. The Project had taken no initiative to reach out to the Department of Health's nurses, present in East Harlem's public schools within the Project's own jurisdiction and at the baby milk stations in the larger East Harlem neighborhood. But, aware that the Rockefeller Foundation had hoped that the Project would unify both private and public health nurses, it did add its half-hearted hope to work toward a more fully integrated nursing service with the Department of Health.[51] But its real aspiration in moving forward was to fundamentally change the way nurses thought about their patients and how they taught their families. Rather than thinking only about the health content needed, nurses now needed to consider the context in which the content would be delivered. They wanted to engage public health nursing practice more deeply in the emerging mental health and mental hygiene movement. More specifically, they hoped to use ideas borrowed from mental hygiene to think about the personalities of those receiving their messages; understand the attitudes that existed among members of the family of which the individual was but one part; and know the "desirable" and "undesirable" traits that might affect the lives of the mother and child at present and in the future.[52] Then nurses could begin their health teaching.

The Project then envisioned another new goal moving forward: to forge a more permanent and formal relationship with TC for postgraduate education for public health nursing leadership.[53] It fell to Folks to convince Ruml of the wisdom of this expanded vision of the Nursing Project as a service and a teaching site. Not only had the Project provided excellent, efficient, and more expansive services, Folks argued, it had also served the Foundation well as a training site for the nursing fellows it selected for advanced training from its sites around the world. Nurse fellows from Japan, China, the Philippines, and central and southern Europe trained briefly in the Project and experienced the best practices in public health nursing that they could incorporate into their own nursing once home. It now wanted Ruml to fund the work with Teachers College to establish a formal Institute of Nursing Education for graduate nurses at the Demonstration.[54]

The Rockefeller Foundation balked.[55] Richard M. Pearce, a noted pathologist recruited from the University of Pennsylvania to become the Foundation's director of medical education for the International Health Division (IHD), had been watching developments in East Harlem with increasing alarm. Pearce's division oversaw the development of nursing as well as medical education in countries in which the Foundation supported public health development projects. He worked closely with Beard, who identified the women chosen for fellowships to study US public health nursing practices; and as early as 1926 she had mentioned to him that a request for additional support for the Nursing Project seemed "inevitable." Pearce, careful to acknowledge that he had no authority over the Laura Spelman Rockefeller Memorial's policies, constructed a memo to his Foundation colleagues clearly outlining what he believed the IHD's position should be. It had no interest in either the Health Center or the Nursing Project. Granted, he conceded, its role in training public health nurses did involve the educational initiatives supported by the Division of Medical Education. But it was completely unrelated to the Foundation's main objective: undergraduate training for public health nurses. The Nursing Project would propose Foundation support of a graduate program, and stand in direct contradiction to its practice of supporting only pre-licensure nurse training schools associated with teaching hospitals of medical schools.[56]

Nursing and the Rockefeller Foundation

The Nursing Project's request came at a turning point in the Foundation's nursing policy. The Foundation had always been clear that its support of nursing was directly connected to its support of medical education and public health, both in the United States and abroad.[57] From its initial work on hookworm control in the early-twentieth-century American South, the Foundation had developed global programs in medical education, research, and public health. Its commitment to help rebuild the public health infrastructure of war-torn Europe crystallized what, for the Foundation, was the critical issue related to public health nursing: What kind of education did a public health nurse need for effective practice? It had already commissioned a report on the educational needs of US public health nurses before the war, a report that expanded to include the totality of nursing education in the immediate postwar period. The Foundation subsequently commissioned a second study on those of European nurses in 1921 under the direction of Elizabeth Crowell, the Foundation's nursing representative in Europe.

In the early 1920s, the Foundation had found itself frustrated that there was little clarity or consensus among leading American, Canadian, and British educators about how to train public health nurses or, indeed, nurses. And Crowell frequently found herself at odds with American nursing leaders. Nursing schools in the United States, she wrote in a 1922 letter to George Vincent, the Foundation's president, were too caught up in the web of a professionalizing agenda to provide a model of the kinds of intensive and personalized care that hospitalized patients on the Continent needed. On the other hand, she continued, the rigorous emphasis on higher education and close supervision found in the United States translated perfectly to a robust public health nursing model that would broaden the scope and the practice of the science of public health in both urban and rural areas throughout Europe.

Crowell, although American trained, understood that she was taking a position that seemed like "rank heresy" to her colleagues in the United States. She preferred the English approach. Its hospital-based training schools, run under the stern guidance of long-serving matrons, instilled both the "spirit of service and the conception of the fundamental, therapeutic value of hygiene, diet, and comfort" in the preparation of a nurse committed to a hospital-based career.[58] As importantly, she remained impressed by an English public health system that allowed for the flexibility of more than just trained nurses engaged in health work. Crowell pressed for England's use of other women in health work, including midwives and lay "health visitors" teaching well families in their communities the basic tenets of good hygiene, diet, and comfort in ways that reflected the different customs and details of their lives.

Moreover, as she traveled through Europe, Crowell remained consistently impressed with the varieties of models she observed for training nurses. She remained particularly struck by France's "Strasbourg Plan." This model explicitly addressed the frequently occurring overlaps between health and social welfare work. It had a core curriculum for both nursing and social welfare students for their first two years of training, followed by a third year of more specialized content that emphasized one or the other particular area of practice.[59] Crowell, in her 1923 report and in all her communications with the Foundation,

supported a plurality of training sites and models adapted both to the particular scope of Foundation initiatives and to variations in educational standards
that were sensitive to the long-existing traditions, prejudices, and politics in
particular countries.[60] This was anathema to American nursing leaders who
insisted on a single standard for global nursing education.

The survey on American nursing education officially published in 1923 as
Nursing and Nursing Education in the United States represented a victory for
American nursing leaders, particularly those leaders in public health nursing.
Given the current state of education for public health nursing practice, *Nursing
and Nursing Education* recommended that all agencies hire only public health
nurses who were fully trained nurses with a postgraduate education in public
health that included both course- and fieldwork. It did acknowledge different
European models of education for public health nursing practice, but it pro-
nounced itself to be "convinced that the teacher of hygiene should be equipped
with no less rigorous training than the bedside nurse, further supplemented by
special studies along the lines of public health and social service." It cited Elizabeth Fox, director of the Bureau of Nursing for the American Red Cross, on the
importance of visiting nurses' entrée into families during times of illness that
built the trust necessary to return to those families and provide health teaching. "We seem to think," Fox wrote to Goldmark, "that our American people
are most anxious for advice." Most public health nurses would disagree, she
continued. "American people think they know how to run their own affairs . . .
and are not anxious to be told by someone else how to do it." Rather, when
the one who nurses them when they are sick and suffering offers advice and
suggestions "they are going to take her advice, because it is . . . counsel from a
person who has helped them in times of need."[61]

Nursing and Nursing Education also looked to the future. It recommended
that generalization be the standard model of public health nursing practice.
It also recommended that hospital training schools rework their curricula to
reconfigure their thirty-six-month curricula. Henceforth, it argued, a high school
graduate with a twenty-eight-month curriculum that emphasized the care of the
sick would then have an additional eight months to learn and to practice as a
public health nurse. This eight-month frame was not arbitrarily chosen: it was
the average length of many of the postgraduate programs then in existence.
But by moving postgraduate education into the pre-licensure training, nurses
would be better prepared to enter directly into public health nursing.[62]

But *Nursing and Nursing Education* was not a complete victory. Its recommendation of a shortened pre-licensure course remained one of the most
"hotly debated" topics. Critics included Annie Goodrich, the new dean of

the Rockefeller endowed School of Nursing at Yale University, who strongly believed that every public health patient deserved a "fully trained nurse" with strong postgraduate training. Goodrich, of course, believed the curriculum she designed for the Yale School of Nursing produced just such a nurse: Her students came to Yale with two years of college education and had both class and clinical experiences with Amelia Grant, a new assistant professor who also directed the nursing service at the New Haven Dispensary. But as hospitals controlled most other training schools, their insatiable need for staffing would inevitably compromise any attempt to include public health content and field experience. And she had experienced this. Early in her career, as the director of Saint Luke's Hospital School of Nursing in New York City, she had carved out space to learn and to practice "social service nursing," an experience that would send the hospital's most talented students into the homes of its discharged patients for continued care and health teaching. This program quickly collapsed when Saint Luke's Hospital added new beds.[63]

None of the report's recommendations, of course, came as a surprise to American nursing leaders. Nurses Mary Beard, Lillian Clayton, Annie Goodrich, Adelaide Nutting, and Lillian Wald sat on its advisory committee; as did physician supporters such as C.-E. A. Winslow, a chair of Yale's Department of Public Health and a strong proponent of public health nursing, as well as Livingston Farrand and Hermann Biggs. And most of these men and women were strong supporters of the East Harlem Nursing and Health Demonstration Project that was slowly beginning to establish a presence in the postgraduate education of public health nurses. But the clear rift among its nursing advisors about determining the way forward for nursing, in general, and public health nursing, in particular, in the United States and on the Continent worried Foundation officials. Edwin Embree, still in charge of the Foundation's nursing portfolio, tried to broker a compromise in 1925 by sending four leading US and Canadian nurses to Europe to survey the conditions of nursing education. Goodrich and Clayton represented the United States. Kathleen Russell, the dean of the University of Toronto School of Nursing, a school that was among the Foundation's favorites because of its undergraduate attention to public health, and her assistant Jean Gunn, represented Canada.

Of course, these three constituencies almost immediately clashed over Crowell's choice of two young Czechoslovakian physicians for Rockefeller nursing fellowships. First, there was the very obvious concern about the selection of physicians for prestigious nursing fellowships. Then, there was Crowell's wish to send them to Yale, where "they would be impressed by the dignity of nursing as a profession and with the fact that the nursing students at Yale would be on the same level, intellectually, as would be their sister students in the medical school." Goodrich was appalled that Crowell thought their training as physicians could lead to a shortened period of nurses training. Russell was aghast: She believed that it would be so much for the better to send these students somewhere else as they "ought not be wrapped up in cotton wool, but ought to be made to see the gaff and see hospital nursing as it exists in 99 of 100 institutions." Crowell won that battle. Goodrich accepted these two physicians at Yale under the threat of sending them to Toronto.[64]

But if she won that battle, Crowell lost the war. Russell and Gunn supported Crowell's position about flexibility in the models for nursing education.[65] But Goodrich and Clayton strongly pushed a globalized American model. In a September 1925 meeting with Embree after they returned, they did praise Crowell for what she had accomplished with limited resources. But they felt that the time had come to insist on higher standards for those nursing in Foundation-supported hospitals. Europe, Goodrich and Clayton argued, would develop moderately good schools of nursing on its own. The Foundation's role should be "blazing trails that later would be generally followed."[66] And almost to the day, concerns arose within the IHD about whether the standards for global nursing education were "sufficiently high." As Frederick Russell, its director, pointed out, the IHD insisted on four to six years of training for public health officers from abroad, and the Division of Medical Education, supporting national fellowships, required "thorough" premedical and medical work. Yet, the Foundation only required one year of training for nurses in Rockefeller-supported European projects.[67] In October 1925, Vincent called Crowell to New York for a series of conferences to settle the "nursing policy" of the Foundation. But, in fact, it had already been established. Henceforth, the Foundation would only support those nurses and nurse training schools that served as "light-houses" that blazed the American trail.[68] Those European nurses chosen by the Foundation for fellowships in the United States now needed "weeding out" in more developed training schools in England or on the Continent where their leadership abilities and technical skills could be demonstrated.[69] The successful candidates could then come as fellows to study nursing education at Teachers College in New York City, nursing practice at the University of Toronto and with the East Harlem Nursing Project, and rural public health practice at the Foundation-supported Vanderbilt University in Tennessee.

Yet the Foundation was beginning to worry about its own "light-houses" in the United States. It hoped that support for collegiate nursing education at Yale, Toronto, and Vanderbilt would create new curricula and training models that would graduate fully functioning public health nurses in as little as two years at the pre-licensure level. It expected that these "progressive schools" would change fundamental undergraduate nursing courses in ways that emphasized public health as well as bedside nursing practices. Instead, these schools, especially Goodrich's Yale, seemed to the Foundation's frustration more akin to "protected schools" in that its students were only relieved of some small part of service obligations on hospital wards and graduated as inadequately prepared to function as fully trained public health nurses.[70] Left unstated was the opinion of such educators as Goodrich and Clayton that public health nurses needed to be—first and foremost—fully trained nurses exposed to all areas of nursing practice: nutrition and medical, surgical, obstetrical, pediatrics, and mental health nursing. Their model replicated that of medicine: a traditional four years of medical school followed by postgraduate training in newly fashioned and research-intensive Schools of Public Health.[71] The American nurses had won their war.

By 1927 it had also become evident that the Foundation's administrative structure was too unwieldy. The Rockefeller Institute for Medical Research, and four Rockefeller Boards: the Rockefeller Foundation; the General Education Board; the International Education Board; and the Laura Spellman Rockefeller Memorial seemed to outsiders unrelated, independent, and equally available for grants. And within the Foundation, too many administrative structures created what its officials believed to be a "twilight zone" into which applications that were not obviously within the domain of the humanities or the natural and social sciences might disappear without adequate consideration.[72] In 1929, the reorganization was legally official. The Foundation now had two boards: the Rockefeller Foundation, which now included the IHD, and the General Education Board. The Memorial was dissolved.[73] Nursing initiatives now lay within the purview of the Foundation, with those involving US proposals under the direction of Thomas B. Appleget, one of its vice presidents, and those involving international ones under Pearce. Pearce, as unhappy with the direction of Foundation-funded nursing initiatives abroad as he was at East Harlem, had already ordered a complete review. Pearce asked Crowell to conduct this review. And remember, he warned her in her letter of instruction, the Foundation's interest in public health nursing education remained at the undergraduate level; responsibility for graduate nursing education, in contrast, lay with the government.[74]

A New Approach to Nursing

In 1927, the still existing Laura Spelman Rockefeller Memorial remained strongly supportive of the actual work of the East Harlem Nursing Project. Moreover, concerns that the pending reorganization might constitute a "public

relations disaster" if no provisions were made for the kinds of charitable philanthropy embodied in the traditional Memorial grants strengthened the Project's argument for another five-year grant to continue its service mission.[75] But
despite East Harlem supporters' resolute claims that the service and teaching
missions were "inseparable," the Foundation refused to move in support of
graduate public health nursing education that the Project's teaching service
represented. To do so would not only be to contradict its stated policies, but
it would be an admission of the dream that someone, somewhere, somehow
could create a real undergraduate school of public health nursing.

Bailey Burritt turned instead to the Milbank Memorial Fund. As he wrote
John Kingsbury in 1928, the public health nursing teaching that occurred within
the Nursing Project had not been part of its original design but rather had been
"pressed upon it" from sources of "responsibility and influence." The plan to
fund a formal teaching service represented a "great opportunity" to influence
the direction of public health practice.[76] This opportunity of influence proved
tempting. The Fund's own Bellevue-Yorkville Demonstration was floundering.
While Burritt had initially believed that the city would eventually capitulate to
plans for Bellevue-Yorkville given that the private Fund had the advantage of
time and could wait for changes in public administrations, by the later 1920s
he had grown increasingly pessimistic. The health and welfare agencies operating in the district of Bellevue-Yorkville, he decided in 1928, were much less
adaptable to political and social pressures than were those in East Harlem.[77]
But neither he nor the Fund were ready to abandon the demonstration. They
had recognized that its nurses needed more training if they were to be successful in its planned door-to-door campaign to convince parents, teachers, and
key community leaders of the value of providing diphtheria immunizations to
their children—seen as a substantive contribution to the city's success in its
campaign to eliminate this deadly childhood disease.[78]

The Fund had recruited Amelia Grant from Yale in 1926 to direct these
campaigns as part of a generalized public health nursing service. Yet in 1928, at
the same time the Fund agreed to support the teaching service at East Harlem,
Grant left the Bellevue-Yorkville demonstration to assume the position as director of the new Bureau of Nursing of the Department of Health of New York City.
For the first time in the department's history, all public health nurses would
now report to their own nursing director rather than, as in the past and as was
typical of most large urban departments of health, to the medical directors of
the various bureaus in which they worked. Grant's new position and responsibilities were of such import to the field of public health nursing that Lillian
Wald made the announcement in the pages of *Public Health Nurse* herself.

Wald described this appointment as "almost without precedent" and as the capstone of a "long deferred wish of pioneer public health nurses."[79] Indeed, members of the Board of Managers of the Bellevue-Yorkville Demonstration, reflecting on their work in 1933, felt that bringing Grant to New York City and then letting her go to the Department of Health had been, perhaps, "the most outstanding contribution of the Demonstration."[80]

The Nursing Project, now secure in another four years of support from the soon-to-be-dissolved Memorial, the commitment of the Milbank Memorial Fund to its teaching service, and the resources of the four cooperating nursing agencies, set about to create a formal "family nursing service" that would represent a "new approach to health work" by more fully integrating knowledge from nutrition and mental hygiene into their work. As the representative agency in East Harlem for the HSS Visiting Nurse Service and the Maternity Center Association, it would use the "medical-nursing approach" of these "covering" services as the basis to build the "relationships for the educational work that would continue long beyond the acute need for the initial service."[81] It would develop a "common program of health education" that would be carried into the home by one nurse whose relationship with the family would continue over time.[82] In 1928, the Nursing Project brought the demonstration part of its work to a close, and reopened as the East Harlem Nursing and Health Service.

Chapter 3

Practicing Nursing Knowledge

By 1931, supporters of the new Nursing Service had a consistent message that it sent to the Rockefeller Foundation in support of its practice and teaching missions. Grace Anderson, in her report to the Foundation on the work of its Teaching Service, spoke directly to its significant success in "pooling of professional knowledge and skills in working out the essentials of a family health program for the community." Only in East Harlem—and, she argued, nowhere else in the country—could observation and practice be directly correlated with theoretical instruction in education, psychology, sociology, nutrition, mental hygiene, and social casework. Its students from around the globe learned about family relationships in class and focused on improving them in practice. It provided its students with a "social laboratory" in which experiences were translated into new principles and practices.[1] And rather than reporting their prenatal and health work with mothers, infants, and children as separate categories, Anderson spoke more directly to their work with families as a whole.

Homer Folks also carried a similar message to the Foundation. As he wrote Thomas B. Appleget, the Foundation vice president to whom the Service now reported, it was now a successful "family service." Its success lay in its specific recognition "that public health nursing is a complex undertaking which must derive many of its techniques from specialists in other fields." The teaching and supervisory staff at the Service now included nutritionists, mental hygiene specialists, social workers, teachers, and physicians. Mindful of the Foundation's concern about where and how a public health nurse should be educated, he also noted the strategic position of a fully trained nurse. Through calls from mothers seeking prenatal care for themselves or home nursing for sick children, such a nurse reached "a cross-section of the community—families that would not be known to other agencies."[2] He reemphasized this in 1932, calling attention to the increasing interdisciplinary nature of the family service. "The Nursing and Health Service has disregarded the barriers that exist between professional groups," he wrote, "and has brought experts in nutrition work, in mental hygiene, in social work, and in education into a close working relationship with nurses and physicians to the end that a more complete service may be rendered to the people of the community."[3]

On one level, this chapter explores the knowledge needed for this reworked notion of public health nursing practice. Some, such as the knowledge required for generalized public health nursing practice, had long fallen within nursing's domain. Other kinds involved knowledge relocations as messages about health and illness became more normalized and standardized. Supported by additional funding from the Milbank Memorial Fund, for example, the Bellevue-Yorkville Demonstration Project charged two public health nurses with developing health education curricular materials that the city's public and parochial school teachers would incorporate into their own lesson plans, freeing up time for these schools' own nurses to incorporate vision tests, formerly the purview of physicians, into their own practices. And still others involved incorporating new knowledge, particularly that associated with the mental hygiene movement, into extant disciplinary practices.

But this chapter is about more than the knowledge required for health work. It is also about how ideas about health circulated between and among constituents, how they were implemented, and how their implementation fed back into new policies and practices. At the Bellevue-Yorkville Demonstration Project, for example, the relationships were fairly straightforward. In conjunction with the Department of Health, it had also prioritized health initiatives, particularly those promoting the periodic medical exams. It hoped its medically rich environment would provide the support and the resources necessary for this campaign. The Bellevue-Yorkville Project fought hard: It invited local private practitioners to the center to learn about and practice this new medical procedure; it sent nurses into their offices to educate their patients; and it offered laboratory services for specimen analyses that were part of a comprehensive health exam. The Project, however, failed: Physicians remained skeptical about a practice for which they had received no training in medical school; and patients remained suspicious that this was just another way for physicians to extort more fees.[4]

At the East Harlem Nursing and Health Service, however, the relationships were more complicated. These nurses, like other progressive urban colleagues

throughout the country, used their practice experiences to move to legitimiz-
ing their claims to families as their exclusive domain. They built knowledge
that bridged the biological sciences that supported their public health practices
with the new knowledge in the social sciences that buttressed their work with
families. This practice, however, brought them out of bounded disciplinary
interests and into a place at the center of not only their own but also others'
agendas. Foundations, families, physicians, and other public health workers
all had particular ideas about what nurses should and could do as they delivered their messages of health. Indeed, the Service's nurses practiced in a very
complicated space of ideas, practice, action, and actors. It locates the problems
of coordination within disciplinary tensions as nurses and social workers—
working within a web of gender, class, race, and power—sought to advance
their own disciplinary interests even as they searched for better ways to care
for the families in their charge. The knowledge they needed for practice was
contingent, determined not just by the needs of its and other disciplines but
also by the demands of the community it sought to serve.

Knowledge for Practice

In 1926, the Nursing Project formally published its research on the comparative effectiveness and costs of generalized and specialized public health nursing services. This pamphlet also included an appendix that described a six- to
eight-week period of staff orientation to and education for generalized nursing
practice where one nurse attended to all the health and illness needs of a defined
neighborhood. But in 1926 and carried through to its 1928 reinvention as the
East Harlem Nursing and Health Service, it could hire what many other public
health nursing agencies could not—it could choose among experienced white
public health nurses.[5] The Service's silence, however, on the backgrounds of
those white nurses new to its practice does speak to its privileged place within
the city's public health nursing community—and, indeed, the very privilege of
whiteness within its tight circle. East Harlem worked within an assumption of
white competence.

Black public health nurses could not. A 1928 press release generated by
the Association for Improving the Conditions of the Poor (AICP) about the
black-nurse-managed Health Center at Columbus Hill needed to carefully elu-
cidate these women's impressive backgrounds. Their supervisor, Sadie Stewart Hobday, first trained as a teacher at the Hampton Institute, then attended
the Lincoln Hospital Training School in New York City, and practiced public
health nursing in Tulsa, Oklahoma, where she wrote about building the city's
own black health center for *Public Health Nurse.* In 1927, Hobday had returned

to New York City where she worked with four other nurses who shared con-
nections to the Lincoln Hospital Training School, the black Harlem Hospital,
and postgraduate training at the Henry Street Visiting Nurse Service. Education
stood as a proxy for class and class mattered for respectable practice, irrespective of race.[6]

But within the segregated norms of the city's public health practices, class,
race, and gender continued to intersect in complicated ways. Both the East
Harlem Nursing Service and Columbus Hill Health Center used volunteers, but
those at the Service were women from the community who helped the nurses
navigate issues of language and customs among its Italian American families.
Volunteers at Columbus Hill, by contrast, were not from its British West Indian
community. Rather, they were married graduate nurses who certainly would
have added valued services. But most importantly, their presence enacted
norms of middle-class black domestic respectability in which wives did not
need to work for money to a poorer community in which mothers—too often,
in their nurses' eyes, single mothers—had to work to support their children.[7]
That these nurses were women mattered as well. Only black nurses could represent black middle-class respectability. The AICP, working through Columbus
Hill, would not hire any formally trained black social workers, preferring to use
untrained black "field workers" and "visitors" to do its social welfare work.[8]

Whether at East Harlem or in Columbus Hill, whether for nurses new to
public health nursing or moving from specialty to generalized practice, the
East Harlem Nursing and Health Service recommended that knowledge for
generalized nursing practice began with a study of the families in a particular
community. During their first week, nurses new to East Harlem learned about
the Nursing Service's mission, wrote their impressions of the community, and
began practicing with nurses, answering calls to provide bedside nursing care
to sick individuals. This seemed the easiest way to draw on the knowledge
that hospital-trained nurses already knew and to move them toward two goals:
beginning to think about individuals in the context of their families; and integrating the knowledge that brought nutritional, mental hygiene, and tuberculosis care into their practices. By 1928, Grace Anderson had also added statistical
knowledge to the public health nurse's repertoire. This kind of knowledge, she
argued, helped the nurse find and, more importantly, interpret data circulating
about the health of their communities. Remember, she cautioned, to use care
when interpreting trends in morbidity and mortality rates; it would be quite
"dangerous" to draw conclusions about particular conditions that included
only a small number of cases since clinically insignificant variations could dra-
matically skew results.[9]

Data collected and systemized by public health nurses, in fact, drove many public health databases. And, Anderson admitted, the extensive, if not exhaustive, systems of recordkeeping in public health nursing practices remained "a much debated point." Statistics, favored by public health reformers and demanded by the philanthropies that supported New York City's health demonstration projects, existed at an uneasy intersection of knowledge and perception. Matthias Nicoll, the commissioner of New York State's Department of Health, advised the Milbank Memorial Fund in 1924 to remember when thinking about the outcomes of its three demonstration projects in the state, that "statistics don't demonstrate." Neither do they "have any effect at all when it comes to a consideration of what that means to the average man in taxes." In fact, Nicoll concluded, "I think he is going to look at his tax bill and take his chance on death."[10]

Similarly, Anderson's nurses looked at the enormous amount of time and energy that went into creating and maintaining data and wanted to take their chance to have more time with patients. They had to create individual files that were cross-indexed with family files. They had to complete separate forms for their maternity visits as well as visits to preschool children, sick individuals, and those patients with tuberculosis. At the end of each day, they had to create their own daily reports on home visits and those in the clinics they conducted at the Service. These data, in turn, fed forms for monthly reports that fed forms for quarterly and annual ones.[11] Anderson's nesting of statistics as important new and scientific knowledge for public health nursing practice hoped to reframe this tedium. It followed a long tradition of reimaging practice through the lens of knowledge. "How hopelessly dull, not to say irritating," Isabel Hampton Robb, a leading training school superintendent of nurses, had admitted in 1903, "would be the many washings and various aseptic precautions which are now required from the nurse . . . unless she had learned from bacteriology to appreciate the fact that there exists a surgical, microscopic cleanliness."[12] A quarter century later, Anderson wondered, is not our direct care of patients so much better when driven by data rather than subjective impressions?[13]

Data also drove the next two weeks of training in maternity nursing: data on the mothers themselves, their places of delivery, maternal mortality, infant mortality, and breastfeeding rates. Maternity nursing was also imagined as a practice that would support a fuller transition to family nursing by concentrating on the mother-infant dyad. But the knowledge needed for this practice was not new knowledge, although it may have been new to some nurses. Required readings such as T. W. Galloway's *Love and Marriage* and Carolyn Conant Van Blarcom's *Getting Ready to Be a Mother* were texts already circulating among middle-class wives and mothers. This practice was to be an instance of knowledge transfer, but now from public health nurses to poorer women.[14] This process had limitations. *Love and Marriage* prepared nurses to walk new wives through "conditions" for a successful marriage that included "normal" sexual relationships without any acknowledgment of the social, cultural, and faith traditions that had an equally powerful effect.[15] And East Harlem nurses were quite critical of those traditions in its Italian and Italian American community. They believed that wives were simply passed from homes dominated by fathers to those dominated by their husbands; that they were "handicapped" by too frequent pregnancies; and that their social life was "restricted" to events involving their local churches. The Service's primary goal was to ensure a safe maternity for both mother and newborn infant. But it also took seriously its commitment to "broaden" the mother's "social contacts." To this end, it created prenatal and sewing classes at the Service that provided both educational and recreational resources.[16]

Van Blarkom, a 1901 graduate of the Johns Hopkins Training School for Nurses, became interested in midwifery through her national work on the prevention of blindness: Her own earlier survey of midwifery practices in Europe convinced her that US midwives' failure to use silver nitrate contributed to the place of *ophthalmia neonatorum* as one of the leading causes of blindness in the United States. Her campaign to regulate midwifery practice led to her own position as the first US nurse to also be a licensed midwife and to the founding, with others at Maternity Center Association (MCA), of the first midwifery school at Bellevue Hospital in 1911. Because of her ties to Bellevue and MCA, *Getting Ready* at least acknowledged that mothers crossed socioeconomic (but not race) classes. Often her recommendations on the need for corsets for an expanding abdomen or supports for milk-heavy breasts also contained instructions (with pictures from the MCA's own collections) on how they might be fashioned from materials in one's home.

Van Blarcom did write elsewhere directly for nurses. Her popular *Obstetrical Nursing*, first published in 1922, subsequently went through three editions.[17] But in choosing the lay *Getting Ready to Be a Mother* for the education of its own public health nurses, the leaders of the East Harlem Nursing and Health Service delivered a powerful message of the kinds of knowledge it valued. Public health nurses had long considered themselves and had been considered by others as the "connecting link"—between patients and physicians, between and among institutions, and between scientific knowledge and its implementation in the homes they visited.[18] Now they were to be the "connecting link" between the knowledge easily accessed by middle-class mothers and that needed by poorer ones. *Getting Ready* contained all the standard prenatal instructions that a physician would give his middle-class patient. It discussed the importance of regular exams to measure the growing child; about the problems (and the solutions to those problems) that might be experienced during pregnancy. It spoke to the need for healthy diets, fresh air, rest, dental care, a cheerful and hopeful frame of mind, a safe and sanitary room for a home delivery. And it concluded with the importance of a carefully structured infant routine, built around regularized times to breast-feed in laying the foundation for the development of a strong and independent adult.

East Harlem's plan to orient to and educate for generalized public health nursing practice allowed its nurses, during their fourth week, to work with experienced nurses on home visits that would consolidate their knowledge and techniques, particularly the bag techniques, that differed according to types of cases. Week four, however, did set aside time for a special class on social case-work that would help nurses better understand the social problems some families they had visited experienced. By week five, the nurses had moved on to the care of infants at home.[19] Much of the material they covered overlapped with Van Blarcom's advice, but, increasingly, East Harlem's nurses focused on specific developmental outcomes. This included bowel training at three months, weaning at six months, and bladder control training beginning at any point from six to twelve months. And they noted with pride that many of their babies were out of diapers at nine months.[20] Like Van Blarcom, they both deplored "artificial feeding" as a leading cause of malnutrition and devoted pages to how to instruct mothers in the proper preparation and storage of formulas. They also took the prevention of the development of rickets as a particular issue, concerned that it caused the childhood pneumonias that represented the greatest demand on their bedside nursing service. Rickets, a softening of the growth plates at the end of a child's bones that led to deformities such as bowed legs, had been rampant among poor, urban children at the turn of the twentieth century. But by the early 1920s, researchers had established the value of cod-liver oil and sunlight in its treatment and prevention.[21] And both the AICP and the nurses in East Harlem remained determined to distribute cod-liver oil and to preach the value of play in bright sunlight to all the infants in their charge.[22]

The final week—one that concentrated on working with preschool children—represented the culmination of all that a family nursing service represented. Remember, the nurses learned, that "everything" affects the well-being of these children: the mother and a newborn child; any illnesses in the family; parental employment (or lack thereof); whether children in the home worked to support the family's finances; and the stability of family, particularly marital, relationships.[23] This practice tested all of a public health nurse's

Figure 4.   The Parents Conference Room at the East Harlem Nursing and Health Service
Reprinted with the permission of the Rockefeller Archive Center.

accumulated knowledge of nutrition; normal childhood growth and development; habit training in independence, self-control, and obedience; and of how
best to advise parents in handling their children's temper tantrums and bed-
wetting.[24] It also tested her more traditional public health knowledge in promoting vaccinations and immunizations in families with young children and in
helping parents negotiate various medical and social services as they sought to
correct such identified "defects" as dental caries, infected tonsils and adenoids,
and infections of the ears, eyes, and skin. Many parents seemed to have found
this kind of public health nursing useful. By 1928, Anderson claimed to have
reached 40 percent of East Harlem's preschool children and, over the course of
the demonstration, provided more than four thousand discrete services. Indeed,
she continued, the problem was not in "finding" these children at home with
their mothers; the problem was in "selecting" those children and families who
could most benefit from among the many more who sought its service.[25]

Practicing Family Nursing

At the same time, Anderson also found herself constantly balancing a commitment to generalized nursing with the need to administer clinics organized
around the medical specialties of the physicians who staffed them. Anderson
tried as much as possible to rotate her staff weekly through the Service's six

infant clinics, three preschool clinics, six tuberculosis clinics, and six general medicine clinics to maintain their generalist knowledge base, but issues of timing, expertise, and personal preferences presented constant challenges.[26] But, as she wrote in 1934, it was worth it. "Because of its flexible program, freedom in experimentation, and its long-time contacts with families and individuals," the Service did not need to restrict its mission to only one purpose, as did the city's private visiting nurse agencies who cared for ill individuals in their homes. Nor did it have to deal with the more rigid administrative structures and bureaucracies of the city's Public Health Nursing Bureau. Hence, she concluded: the need. "The Nursing and Health Service offers a type of community service that can only be given by a private or voluntary agency."[27]

Irrespective of the families' wishes, the Service considered itself responsible for all the families it served until they moved out of the district. It had created a new "midway file" for those no longer needing home visits but still needing clinic assistance for routine physical exams or particular tests for suspected cardiac, venereal, or infectious diseases. Advancement to the midway file also spoke to a moral decision made about the family by the Service. Such a parent had assumed responsibility for helping her (and it was almost entirely "her") family; its children were thriving; and she only needed reminders about upcoming classes and clinics.[28] It should be noted, however, that not all families were treated equally. The nurses at East Harlem, believing that the "informed intelligence" of parents was key to their health work, selected young parents with either their first or only a few children and who showed "promise of an ability to learn new ways" for more intensive educational health work.[29] Given the community's demographics, these were likely to be second-generation families.

Families of East Harlem

Its children, however, looking back at their and their families' experiences in East Harlem, seemed much more ambivalent about what the Service wanted to provide. Leonard Covello, an educator deeply involved with Italian American students both in East Harlem high schools and at Columbia University's Casa Italiana, had assigned his college students in the 1930s the task of collecting memories and impressions of East Harlem in earlier years. Covello, who had immigrated to East Harlem in 1896 from Southern Italy as a young child with his family, was a leader in New York City's intercultural education movement who also had an abiding interest in helping his students to simultaneously Americanize while remaining proud of their Italian heritage. The research of his students painted a different picture of East Harlem than the one the public health nurses had internalized.[30]

These informants spoke to a more nuanced sense of community, differentiating the needs and aspirations of those arriving before World War I and those arriving after. Those that came in the early twentieth century had fled the poverty of small towns and farms. They had never experienced urban life. But those that came after the war but before the final immigration caps of 1921 and 1924 had been "dragged by war from place to place" as officers in the Italian army, had "opportunities to improve their minds" and seemed more educated and sophisticated. They were a professional group, more interested in improving neighborhood conditions and much less interested in the "noisy and colorful religious celebrations" and the "antiquated . . . and in most cases useless associations" that had formed the fabric of East Harlem's social life.[31]

Death did seem to pervade their lives, and they spoke movingly of their own experiences of the deaths of parents and of children. "My father's death," one recalled, "were [sic] the periods of my greatest crisis. I thought things would never go straight because the supporter of my family had gone to rest." Yet another remembered that "the worst experience I had as I look back upon them even now is the deaths of my two brothers and mother."[32] They sought consolation in the mutual aid societies they created. East Harlem alone was home to more than 250 such societies. These societies, it seemed, rarely lived up to their promise of help with medical expenses. But they did allow an older generation to place their "health, their lives, and their material fortune in the hands of a benevolent saint without which, for the elders, life would seem impossible." More pragmatically, they also provided for the costs of a "dignified burial," mitigating a "constant fear" that now in the United States a new generation of relatives "might let their traditional duties slip."[33]

Yet concerns about their health did not figure prominently. Most believed themselves to be "sturdy" and "endowed with a strong constitution."[34] They believed in the efficacy of their own or their friends' remedies for "minor illnesses." And they believed in the place of traditional beliefs and practices in the face of scientific evidence. Italian mothers, another informant reported, believed that children, like animals, knew instinctively what is good for them by its taste and would never force them to drink something "strange or disagreeable."[35] The nurses in East Harlem, it seemed, faced a particular challenge when urging mothers to give their children cod-liver oil to prevent rickets.

They faced a similar challenge when urging families to practice healthful habits. As at Bellevue-Yorkville, schools had become an important nexus for the dissemination of health practice education. Yet a "painful contrast" remained. "The teacher said, for instance," one informant remembered, "that clean hands, clean clothing, and a toothbrush are essentials; or that plenty of milk should be taken in the morning." Yet, while not rejecting such messages, inertia set in when trying to translate such instructions into the fabric of family life. "But the father comes home from work, and the mother at her household tasks though they may not oppose the rules, they do not necessarily exemplify them at home."[36]

Families played a seemingly more direct role in decisions to follow the East Harlem nurses' exhortations to seek dental care for their children. One of Covello's students, Alice Kraus, felt the problem of families and dental care was twofold. The first involved the fact that East Harlem families came from rural settings with hard bread as a diet staple. Hard bread seemed to obviate the need for dental care in Italy, but now these families had settled in a city and country "which has one of the highest ratings in the world for dental caries." The second involved the parents' own experiences. These parents—with no tradition of routine dental care—waited to seek treatment "until the pain has practically pushed them into the dental chair." As any good parent, they "feel they are protecting their children against possible pain if they prevent them from attending the (dental) clinic."[37]

Like many ethnic communities, the Italian American community in East Harlem prided itself on its ability to support a hospital for its most destitute members. The Italian Hospital of New York City had initially opened in East Harlem in 1891: It was "small and unpretentious" and its nursing care (indeed all the care, including that of pharmacists and lab technicians) was provided by the Missionary Sisters of the Sacred Heart, an Italian nursing order founded by Francesca (Frances) Cabrini, which also ran hospitals in other cities in the United States with sizable Italian populations. The Italian Hospital soon moved downtown to the Lower East Side, another poor and overcrowded area of the city with a larger Italian population. By 1928, however, when the demonstration project converted to a Nursing Service, plans were in motion to build a new hospital, "an imposing and monumental structure," in East Harlem where patients could find their care delivered "in their mother tongue." But the new hospital would not be for those interviewed in East Harlem: it would be for others. A deep suspicion of hospitals fused with the community's sense of themselves as "sturdy." In the hospital, one 1905 commentator noted, "they say you can't kill an Italian."[38] Yet the Italian American community recognized that a hospital represented more than a way to care for its most destitute members, that it was critically important to its community of Italian American physicians. It meant, as one physician noted, that it provided "incalculable advantages" to a new generation of Italian physicians shut out from practice privileges at the city's more prominent medical institutions.[39]

This deep-seated sense of marginalization pervaded memories. Covello himself described his own feelings of "hurt" at the barriers to full inclusion in the American dream he felt as both a child and a young adult in East Harlem. Although he had later come to understand "that none of these hurts were deliberate"—that they were the unfortunate results of a failure on everyone's part to understand each other—other data suggested he was being unduly charitable.[40] A 1933 series of brief "on the street" surveys of New York City's attitudes toward Italians found them described as "wops," and characterized as "greasy," "dirty," and "destructive."[41] These attitudes found their fullest expression in the East Harlem public schools. At best, one informant remembered, "being Italian was virtually a faux-pas and the genteel American ladies who were our teachers were tactful enough to overlook our error."[42] At its worst, the relationship between students and East Harlem children—particularly its boys—were characterized by mutual contempt. Boys, another recalled, would be deliberately "rebellious" to a female teacher who found it "no small task to come each morning to try to pound ideas into a bunch of little garlic eating greasers." School, he remembered, was conducted in an atmosphere of "bitter opposition and intense conflict."[43]

And schools, Covello wrote, seemed to fail families, as well. Too often, he continued, they refused to give families "even the courtesy that may rightly be expected," and tolerated teachers who "take refuge for their own insufficiency" in disparaging comments about the community.[44] Covello was particularly interested in positioning schools as a "coordinating agency" that might meet a broad array of social, health, and hygiene needs. In his vision, families and schools mutually shared responsibility for and participation in the community's health and well-being. But first, those participating in working toward this particular vision needed to understand "EXISTING ATTITUDES." They needed to recognize that in East Harlem many families opposed taking medicines, were afraid of hospitals, did not fully understand advances in medical science, and must be educated in their own language and in ways that do not "overburden them." But, most importantly, health education had to be dissociated from social welfare and charity work. It had to come into schools where it would instead be identified with "common sense, general education, and the ordinary routines of school, home, and community life."[45]

Covello's vision, of course, fit within the long history of parent education as a social movement in the United States. Its more organized form had begun with the earlier twentieth-century efforts to organize mothers in a common cause around what would become the Parent Teacher Association. It changed during the 1920s as new insights from the social and behavioral sciences fused interests in parent education and the needs of the preschool child. As historian Steven Schlossman has argued, this new movement stressed "the plasticity of early childhood and the irreversibility of character traits instilled then." In keeping with this new knowledge, East Harlem nurses' emphasis on teaching families how to instill regularity and routine into the lives of their infants and preschool children was to prevent the "maladjustments" that would eventually produce emotionally immature and socially irresponsible adults.[46] But, Schlossman continues, by the 1920s, parent education initiatives also moved away from an earlier generation's interest in engaging poor families to focusing on middle-class ones.[47] This left a void that public health nurses and also social workers associated with charity work tried to fill. And neither Leonard Covello nor many families in East Harlem trusted social workers.

Covello was particularly strident. Social workers, he preached, must treat Italians and Italian Americans as "NORMAL HUMAN BEINGS." They must understand the norms and values of traditional Italian families, ones that emphasized the solidarity of the family and the respect shown to elders whose words had the effect of "law." Social workers needed to drop their focus on the "pathological" and understand the "stress" of unemployment, serious illness, and, above all else, their children's transition to school where American values of individualism and choice collided with tradition norms of duty and responsibility to family. Social workers, in the end, needed a "deliberate rearrangement" of their attitudes.[48]

East Harlem families shared Covello's sentiments. They resented "even well-meaning attempts" to help poorer families access material resources because they believed it drew on a wellspring of assumptions that all Italian families were poor.[49] They also believed social workers intrusive. One informant told a story of a social worker who took it upon herself to throw a disabled husband out of his home with a warning that he could not return until he had found employment.[50] And they actively protected their own power. When another East Harlem mother was summoned to a school to explain why her fifteen-year-old son had to work rather than attend classes, she made it known that she had to be present at all follow-up interviews. As she explained, she would not have her son "exposed to the influence of the social worker without the benefit of his mother's judgment."[51]

Social workers lived this legacy of suspicion. Their work in the 1920s demanded intrusion into private family matters as they believed they needed to know the family's history, its current needs, and its potential for future independence through the "adjustments" they would bring to bear.[52] This was the territory that public health nurses sought to enter. At East Harlem, this required not only a change in focus, but also one in methods: Traditional public health nursing checklists had evolved into narrative Family Date Sheets transcribed by stenographers; and health clinics and conferences at the center would become places where a mother would bring all her children at one time rather than sequentially to identified infant or preschool health screenings. But, most of all, it meant more aggressively promoting the mental hygiene aspects of the new family nursing role to a wider audience. Lillian Wald was reportedly "a little afraid" of this change: it would make the nurse "a thorough welfare agent."[53] Or, in other words, it would make a nurse a social worker.

Nursing and Social Work

Public health nurses and social workers had a long and often tangled history. Both groups of predominantly women clinicians came of age in the early twentieth century's Progressive era concerns that the increasingly reductionist and impersonal medical emphasis on the patient's body completely ignored the social and environmental determinants of health and wellness. The response had been the creation of social service departments in larger and more progressive hospitals that would consider the patient as a person with real social, environmental, and economic needs. And, ideally, it would be nurses who would staff such departments. Slowly, the idea of women in social service to the sick and the dispossessed took hold.

The boundaries between the work of nurses and that of social workers remained quite fluid in the early decades of the twentieth century. Indeed, the settlement house movement seemed to completely collapse such boundaries. Lillian Wald, for example, trained as a nurse, but her commitment to reform placed her work within the tradition of social work service. Jane Addams, the head of the equally renowned Hull House in Chicago, had no healthcare background, but her realization of the role that health and illness played in the ability of immigrants to secure a firm foothold in the American experience led to the creation of weekly health clinic services for the residents of her community. Ida B. Cannon, a nurse with experience investigating the living conditions of individuals with tuberculosis in early twentieth-century Boston, established one of the first formal hospital social service departments at the Massachusetts General Hospital.[54]

Yet, Cannon, like many other Progressive-era social workers, felt a compelling need to differentiate her work from that of nurses, on the one hand, and that of religious orders and individuals doing private charity work, on the other. These social workers grounded their work in the newly developing social sciences, established schools of social work and philanthropy, and, with the 1917 publication of Mary E. Richmond's landmark book, *Social Diagnosis*, laid claim to the specialized knowledge and investigative procedures that established "casework" as their unique, professional contribution to the health and well-being of individuals and families.

Like nurses, social workers struggled to get other professionals to recognize their claims.[55] In his famous 1915 address, "Is Social Work a Profession?," Abraham Flexner told social workers assembled at Baltimore's National Conference of Charities and Correction that they were more akin to trained nurses: They were both "twilight cases."[56] Flexner's opinion mattered. He was the author of the influential 1910 Carnegie Foundation report on medical education in the United States and Canada.[57] This report galvanized public and philanthropic support for medical education reform and research; and Flexner was currently serving on the Rockefeller Foundation's General Education Board, setting policies in place that would ultimately transform not only the ways in which physicians were educated but also their place at the apex of social and professional status. In his mind, both social workers and trained nurses, however valuable their work, served in a "mediating" rather than an independent capacity. They both worked through others—physicians for health, legislators for reform—to achieve their ends.[58] The public health nurse, on the other hand, seemed closer to professional status. Flexner believed her to be a "sanitary official, busy in the field largely on her own responsibility rather than in the sick room under orders." Indeed, he wondered whether the title "nurse" appropriately described her role and ventured to predict that term would change as public health nursing matured as a field.[59]

The resurgence of the mental hygiene movement in the 1920s offered a potentially new knowledge base—that of psychoanalytical theory—that both social workers and nurses hoped would buttress their claims to specialized knowledge and independent practice, the two criteria that Flexner believed compromised their claims to professionalism. Mental hygiene, of course, was a diffuse and contradictory term that both promised mental health and strengthened the theoretical foundation of the eugenics movement. But at its core, and in the wake of public alarm about the rates of psychiatric issues identified in the First World War's draft screenings, it promised the prevention of almost hopelessly untreatable mental diseases through strategic behavioral changes such as carefully thought-through techniques of habit formations in children, and intelligent parenting that would provide neither too much nor too little affection.[60] Indeed, the interwar period's shift of the mental hygiene movement from asylum-based psychiatric practices to public health forced nurses and social workers to rethink both their disciplinary practices and their relationships with each other. Social workers, not nurses, had developed the "casework" method for systematically understanding an individual in the context of his or her environment.[61] But nurses, not social workers, had the experience and the expertise in the kinds of neighborhood engagement and outreach necessary for widespread mental health education and treatment.

This "new approach" to public health nursing practice incorporated both the techniques of casework and the nurses' long-standing skills at engaging parents of physically ill children as the best way to reach a child with emotional problems. This put nurses in an acknowledged competition with social workers. In New York City, Harry Hopkins, then with the New York Tuberculosis Association, and Bailey Burritt, the general director of the AICP, conceded physical health and welfare to the East Harlem demonstrations, but they had both hoped that the Bellevue-Yorkville Demonstration would establish the ascendency of social workers in family health work. But, a 1922 editorial in the *American Journal of Public Health* warned social workers that health—both of an individual and of the public—remained the single most important factor in all social work practices, and ill health was the most important cause of poverty and of the "maladjustments" it engendered.[62] This position strengthened the hand of public health nurses who took power from their command of medical science. And nurses used this power. As Elizabeth Anderson, a psychiatrist and consultant to the Bellevue-Yorkville Demonstration, looked back and wondered why cooperation so often failed to take root between public health nurses and social workers, she chided that nurses might remember that they "do not have the final and only right answer."[63]

It fell to Sybil Pease, a social worker and mental hygiene consultant to the Nursing Service, to describe how nurses managed this interdisciplinary new approach to family nursing in a series of journal articles and reports to the Rockefeller Foundation. There was, she acknowledged, the problem of "double identification." A social worker working for a public health nursing agency needed to be able to accept nurses' own use of social casework methods "with equanimity and a minimum urge to change it except as it becomes ready to change itself in its own way and in its own time." As an aside, she added, "in her most optimistic moments she would never have imagined beforehand finding a group in another profession with which she would be so proud and happy to be identified."[64] Pease carefully drew practice boundaries around only a particular group of nurses—public health nurses—who could encroach on the social workers' territory. She mitigated conflict by declaring that the public health nurse was, in fact, a new kind of social worker. Her place in the community stood between caseworkers, who diagnosed and treated the environmental ills that led to serious delinquency and psychiatric illnesses, and settlement house workers, who developed educational and recreational activities for groups in their particular neighborhoods.[65]

It did, Pease acknowledged, take some "working out" to arrive at this role definition. The initial East Harlem Nursing and Health Demonstration Project, under pressure from the National Committee on Mental Hygiene, had first tried to establish a traditional mental hygiene service to meet the needs of the neighborhood. Several afternoons each week a team consisting of a psychiatrist, psychologist, nurse, and social worker would meet with "badly adjusted" individuals and families for diagnosis and ongoing treatment.[66] This proved problematic: The public health nursing staff spent an inordinate amount of time with the identified "problem" child, leaving little time for the community-based family education work also expected of them. By 1930, the Service decided, instead, to concentrate its nurses' work on issues related to mental health promotion and the education of women and, through them, their families. Their research agenda shifted to identifying the knowledge and what they believed to be the social power to be gained from working with well-adjusted families. This included identifying the mental hygiene issues of pregnancy; charting the emotional growth of young people; and developing sex education resources for parents and children.[67] As Pease explained in an accompanying article, East Harlem nurses drew from their long tradition of health education work and reframed their prolonged contact with a mother—from her first prenatal visit through her child's entering school—as a unique opportunity to support mental as well as physical health.[68] Families struggling with issues of severe mental "defects" or disorders would be referred to other social service agencies.

Two additional threads consistent with the mental hygiene movement ran through East Harlem's new approach to family nursing. First, there was no longer any notion of a "normal" family. The pervasive idea of "adjustment" as a signifier of mental health and illness also held it to be but a matter of degree, and that "to be normal is to have a problem of adjustment to work out."[69] All families needed mental hygiene help. In fact, to be "normal" was to be in need of advice "about innumerable things from a friendly person in whom one has confidence."[70] And patients, sometimes termed pupils, would necessarily have that confidence in one who nursed the sick when she returned to tell an expectant mother about infant care feeding and the best weaning practices that would encourage both excellent nourishment and emotional independence.[71]

The second thread consisted of the intense scrutiny expected by the nurse of herself as well as the family in her care. To be more "objective"—to have the capacity to deal with a family's situation without allowing her judgment to be affected by emotions, assumptions, biases, or preconceived notions—she had to constantly examine her own thoughts and feelings. Indeed, the nurse had to be open to the pain as well as the joy of her own emotional life so that she could accept that of others.[72] One important role of the mental hygiene consultant, Pease concluded, was helping the nurse perform a sometimes painful self-examination.[73]

The stakes seemed high. By the early 1930s, Bunds had appeared in German neighborhoods in Bellevue-Yorkville, marching in support of Adolf Hitler; uptown in East Harlem, Fascist rallies and newspapers extolled the leadership of Italy's Benito Mussolini. As Pease said upon concluding her speech to public health nurses in Canada in 1934, the public health nurse directly affects the process of family-building. And in a successful family, "people who have known love and security and a chance to be independent in their first years are not likely to become insane or neurotic as adults; and because happy people do not commit crimes nor does a contented nation make war."[74]

This "new approach" was not easy to learn. As Grace Anderson wrote to Mary Beard in 1936, "the helplessness of many students when deprived of their basic nursing skill is evident." Beyond suggesting attending the Service's clinics, most students were "at a loss as to how to proceed" when in the mothers' homes. In addition, "they expect results to follow quickly on their 'advice.' If the mother accepts advice and results follow, the average student sees nothing beyond an immediate correction. If advice is not accepted, then there is nothing to do because 'I've told her, and she won't do anything about it.'"[75] The challenge was in creating a new worker who understood the "principles" that caused such seeming "non-cooperation," the "reasons" for all behavior that seemed "indifferent, unintelligent, or even vicious on the part of parents," and the ways in which such knowledge could construct a more helpful approach to the family.[76] This would be one who struck a "desirable balance between the passive approach of a social worker and the authoritarian methods of the average nurse."[77] Even in practice, the constant balancing between physical and mental healthcare demanded by the "new approach" proved difficult.

East Harlem nurses were not unique in their need to establish role boundaries with social workers. Between 1931 and 1932, a joint committee of the National Association for Public Health Nursing and the American Association of Psychiatric Social Workers met to consider the role of the mental hygiene consultant in public health nursing agencies. A consensus emerged that the idea of integrated mental hygiene nursing practice rather than a separate mental hygiene clinic allowed for flexibility and the ability of individual agencies to experiment with operationalizing its practice. Ideally, at East Harlem, the nurse would teach content woven from nutrition, medicine, habit training, and education and allow her carefully developed relationship with the mother to carry the message.[78]

Still, the issue of the problematic nature of the relationships among public health nurses and social workers rippled across the East Coast cities fortunate to have relatively large numbers of public health nurses and social workers. A 1928 study of the relationships between public health nurses and social workers in Boston found 15 percent of the cases that required collaboration "problematic." Some involved issues about which little could be done, particularly those involving high levels of turnover among clinicians. Most others, however, involved unresolved tensions between social workers disregarding nurses' intimate knowledge of the families they referred and nurses not appreciating the time social workers had to take to ensure the right kinds of material and emotional relief needed.[79] The year 1929 saw the report of the Committee on Psychiatric Social Work in Public Health Nursing Agencies, commissioned by the American Association of Psychiatric Social Workers, in conjunction with Henry Street's VNS, the AICP, the East Harlem Nursing Service, Chicago's Infant Welfare Society, Minneapolis's VNS, and Boston's Community Health Organization (that city's VNS equivalent). This report confirmed the possibilities and the problems of relationships among nurses and social workers.[80] By 1930, the relationships between both groups were characterized as that of "stepsisters"—alliances of divergent individuals rich in possibilities but dependent upon flexibility; indeed, social workers felt that nurses were "too sentimental" and nurses found social workers too objective.[81] By 1931, Burritt had commissioned a study at the AICP to determine if one worker—trained as part nurse and part social worker—might be a viable alternative. He wanted to again test the waters by introducing the European notion of the "health visitor" into the American healthcare landscape.

The idea of a health visitor had reemerged in 1925, scarcely two years after the Goldmark Report emphasized the need for a fully trained nurse to engage in public health nursing. The British Royal College of Nursing received government approval to establish a standardized training course for health visitors, and local health districts began offering them more secure employment contracts than those received by trained nurses working in the community for philanthropic organizations. This British precedent seemed attractive to the social work reformers leading New York City's two demonstrations projects. Burritt and Kingsbury, in particular, lionized the role Sir Arthur Newsholme, as medical officer of the Local Government Board in Whitehall, played in expanding the role of the state in providing both preventive health services and actual medical care. England had long been viewed as the "birthplace" of modern public

health nursing. But by the early 1930s, when social workers looked at the idea
of health visitors with interest, US nurses looked on, some with trepidation and
others with disdain.[82] Alma Haupt, the director of the AICP's Bureau of Nursing, wondered "Whither Nursing" in a 1929 memo to Burritt. Her nurses, she
wrote, would be happy to leave the AICP's service if the Association provided
better-educated social workers than it presently had.[83] Marguerite Wales dismissed the idea. In England, unlike in the United States, she wrote in the pages
of *Public Health Nurse* in 1930, a public health nurse might mean a graduate
of a specialty institute with impeccable educational credentials, a graduate of a
training school with or without additional midwifery training, a village nurse
midwife with only a few months of hospital training, or a health visitor.[84]

Looking Forward

Little came of this proposal. The Depression hit New York City and the world,
at first slowly and then with blazing speed, bringing to a halt any thoughts of
experimentation or demonstration. Nurses (those that were left after budgets
were slashed in both the Department of Health and private agencies) had to
nurse; social workers (seeing their ranks expanded by less well-trained men
and women) had to determine relief eligibility; and the heads of the most prominent social welfare agencies had to reconfigure their diminished roles as first
the state and then the federal government assumed responsibility for direct
financial relief. The early 1920s nightmares of six or seven workers descending
on one family ceded to those of trying to find just one worker to address crushing needs.

Still, the nurses at the East Harlem Nursing and Health Service remained
resolutely optimistic and secure in their new knowledge claims as they looked
to their future. They hoped their claim to interdisciplinarity would eventually
engage the Rockefeller Foundation. The Foundation was, in fact, very interested in interdisciplinarity and in the late 1920s had begun funding an increasingly coherent program of research that focused on the "science of man" and
called for "fearless engineering" to integrate the social sciences with the biological, medical, and natural sciences. The results established the legitimacy of
social science as an academic discipline and eventually lay the foundation for
a new field of science: molecular biology.[85] But, however fearless, this program
of research centered on a particular vision of science that assumed a reductionist stance: that one could isolate and measure discrete variables likely to have
the most significant impact. As Ellen Lagemann had argued, this inherently
gendered stance blinded Foundation officials to alternate modes of inquiry
that might focus on more inclusive, comprehensive, and responsive attempts

to employ a more multifaceted approach. They dismissed all other research as "propaganda" by well-meaning but pre-professional (usually) women.[86]

These nurses and their allies also remained convinced they could breech the layers of distrust that East Harlem families held for other workers who would tell them how to live their lives and raise their children. They placed great faith in the premise that the needed bedside care they provided to families in times of illness would translate into families' confidence when they returned to teach the new message of health. They recognized they were not a complete "family nursing service." The nurses in East Harlem, unlike those working for the city's Health Department in the Bellevue-Yorkville demonstration site, had no access to children in their neighborhood schools that were playing an increasingly important role in the circulation of messages of health and hygiene. They nevertheless believed that their approach to teaching mothers about their entire families' health needs would spread their influence even to these children.[87]

As it looked to the future, Anderson and her colleagues remained committed to forging "a new approach" to formal public health nursing practice. But they also looked to take a more visible leadership role in public health nursing education. The nurses at East Harlem still hoped to forge a unique and model relationship with Teachers College at Columbia University for the postgraduate education of diploma-trained nurses who sought public health nursing positions.[88] They also wanted to take curricular leadership in public health nursing and fundamentally change the way nurses thought about their patients and their patients' needs.

Chapter 4

Shuttering the Service

New York City's nurses and social workers witnessed firsthand the devastation wrought by what was at first haltingly described as a "business depression," or an "emergency." The Research Bureau of the city's Welfare Council, the now new and privately funded agency also addressing the issue of coordinating the delivery of health and social welfare services, turned to them to gain an initial "impressionistic" view of the plight of families and individuals during the harsh winter of 1930 and 1931. Some nine hundred women making close to their normal four thousand visits to homes each day between October 1930 and May 1931 participated. Most of these women had the kind of long-standing service in their neighborhoods and communities that allowed them to compare conditions before the economic collapse with the current ones. They saw "an unusual and disturbing amount of suffering." They saw cases of "actual destitution," and families on the brink of such destitution because families and friends they might normally rely upon were in similar circumstances.[1] Social workers felt grim. They experienced dramatic increases in cases, often marked by what they believed to be expectations for financial assistance rather than more humble and embarrassed requests for any resources that might be forthcoming.[2]

The city's public health nurses, such as those working in the schools in the Bellevue-Yorkville Demonstration, noted difficulty in evoking interest among parents in their messages of health education for their children. They remained sympathetic, noting that messages about, for example, dental hygiene, would not become a priority among parents "who were worried about the rent and the next meal."[3] Those at the Henry Street Settlement's Visiting Nurse Society remained adamant about separating their bedside nursing from any kind of relief work. Unlike groups such as teachers and police officers who had devised impromptu relief strategies, they believed, echoing Wald's need to maintain the boundary between nursing and social welfare, that such work would be distracting from its core mission of nursing the sick.[4] But, overall, public health nurses in both public and private agencies felt cause for optimism about the long-term effects of the economic crisis on the future of their discipline. Perhaps, they wondered, "the time and attention they gave to helping people about their economic problems in time of need may bear fruit in a greater willingness to heed the advice of a nurse when she goes to them about matters of health."[5]

Nurses at the East Harlem Nursing and Health Service knew of the effects of this massive economic collapse. The Depression and the accompanying unemployment had hit the community early and hard. An informal survey of families receiving its services in January 1934 found 73 percent dependent on outside sources of income; 24 percent with a bare subsistence income; and a mere 1 percent as moderately comfortable. A more formal 1934 survey of 602 families found 61 percent of families on relief; and of the 37 percent still described as "self-supporting," 22 percent were still vulnerable as they were living on savings or with other family members.[6] Yet, East Harlem nurses also felt reason for optimism. "In the face of the depression conditions, these families have maintained their morale and their children's health to an amazing degree."[7]

But if the East Harlem nurses knew about their families' economic vulnerability, they thought little of the changing social and healthcare landscape. Throughout the 1930s, Puerto Rican families increasingly settled in the neighborhoods of East Harlem. Moreover, these families were moving into a public healthcare system increasingly dominated by the rise in hospitals and outpatient clinics where families increasingly sought medical care. This chapter argues that the nurses in New York City's demonstration projects paid little attention to warnings about the implications of these new clinical sites for public health practice. They steadfastly maintained the site of their practices to that place where they thought it could be most effectively and independently exercised: with cooperative families in their own homes, in the clinics the nurses controlled, and in the classrooms they created. Despite their commitment to maternal-child health initiatives, this narrow focus allowed them to ignore professionally one of the most pressing public health issues in the city in the early 1930s: the newly rising rates of maternal mortality attributed by both the New York Academy of Medicine and the Maternity Center Association to poor obstetrical practices in hospitals that women were increasingly choosing as sites of their infants' births. These nurses could not see or take responsibility for solving problems that lay inside public health policies but outside their defined disciplinary purviews and sites of practice.

A Changing Landscape

In many respects, East Harlem Nursing and Health Service faced changes over which it had little control. First, the Service had no permanent home since the deed to the American Red Cross Building that housed the East Harlem Health Demonstration Project passed to the city when it finally came to an end in 1932. The Service maintained a temporary residence in the building, but it was not until 1934 that a philanthropist deeded to them the building that housed the short-lived Kips Bay Day Nursery that she had supported. While grateful to have a final home, that building lay in the northeast rather than the central section of the district, and attendance at the Service's clinics dropped by 20 percent. Given the travel distances involved, the nurses advised parents who lived in the southernmost section of East Harlem to register their children at the city's Baby Health Station serving that area. The Service now served a smaller community of 57,000 individuals. But it claimed that its "family health service" reached 63 percent of the community's newborns; 40 percent of its preschool children; and 21 percent of its maternity cases. More telling, however, was a Service that had always prided itself on meeting all calls for sick nursing in homes now reported without comment that it met only 34 percent of neighborhood's need for "traditional bedside nursing."[8] The Depression had certainly begun to take its toll. But it had also provided a subtext for the Service's move from more traditional and labor-intensive practices where private public health nursing had been in the past to the focus on the public health teaching where it wanted to be in the future.

Also, the Depression had rocked private, voluntary agencies that could not meet overwhelming and legitimate needs for economic relief. In a complete reversal of numbers attending its opening in 1922, 98 percent of East Harlem families needing relief were now supported by state and federal agencies; only 2 percent received support from private agencies. Until the 1930s, these social workers had been employed by the private agencies that had long supported the material needs of East Harlem families nursed by the Service. Now, they too faced crushing caseloads, dwindling resources, and a steady erosion of the time needed for the personalized, face-to-face interactions demanded by their casework method. Those few who could, moved into private, fee-based marriage and family counseling practices. The many who could not either found themselves

Figure 5.   The Children's Play Room at the East Harlem Nursing and Health Service
Reprinted with the permission of the Rockefeller Archive Center.

unemployed and eligible for relief, or joined the increasingly bureaucratic arrangements that first city, then state, and then federal relief programs needed for their administration.[9] These arrangements were now highly gendered. As a 1935 Welfare Council report noted, "women now occupy the great mass of the poorly paid positions upon which the social work structure rests."[10] And they were fraught with distrust. As historian Daniel Walkowitz has argued, many political opponents of publicly funded relief programs remained profoundly suspicious of the true needs of those deemed eligible. The city's own internal memos reminded its social workers to limit their investigations to eligibility determinations; it was "not . . . to deal with personal and family problems."[11] On the one hand, the East Harlem Nursing Service could now lay sole claim to the disciplinary prerogative of family nursing.

Yet, on the other hand, those same federal dollars undercut their community focus. Fiorello LaGuardia, a child of East Harlem and now the mayor of the city, was committed both to public health (his first wife and their child died of tuberculosis) and to the new federal construction dollars available through the Work Projects Administration. Under his watch, the city secured millions of dollars to expand dental screening programs, provide preschool health exams, add public health nurses to the Health Department rosters, to build hospitals, and, with the full support of the Welfare Council, to bring a neighborhood health center to twenty identified districts in the city.[12]

Neighborhood Health Centers

The roots of this return to neighborhood health centers lay in the same impulses and the same men that supported the city's health demonstrations in East Harlem and Bellevue-Yorkville in the earlier 1920s. But this was a broader approach to a city in which there existed a "wide gap" between those working in health fields and those interested in community development. These gaps were geographical: the reach of hospitals, dispensaries, settlement houses, community centers, and neighborhood associations cut across the Department of Health's division of the city into "sanitary areas" coterminous with census tracks, leading to confusion as to which organizations provided which services to what neighborhoods. These gaps were also about authority, class, and ethnicity. As John Gebhart of the Association for Improving the Conditions of the Poor (AICP) had previously argued in a 1923 meeting, it also involved the kinds of expert knowledge needed for effective action. "Lay interference" from members of the community itself, he announced, "unreasonably delayed or frustrated" expert judgment on action needed.[13]

The appointment of Shirley Wynne as the city's reform-minded health commissioner in 1928 spurred new interest in broadening the health center movement. Wynne appointed a formal Committee on Neighborhood Health Development in 1929, convinced that the delivery of health services should be nested in not only the needs and but also the voices of the particular neighborhoods it served. But his tone, as befit the politics of his position, was different than Gebhart's earlier one. In his vision, such neighborhood health centers represented the "democracy of public health" that would make living in a congested, complex, and at times impersonal city more hospitable. Such health centers, then, would exist as a "living part of the activities of people in the neighborhood."[14]

Wynne had followed the work in East Harlem with interest. It had "demonstrated," he wrote, that bringing together voluntary and public health and welfare agencies prevented duplication and improved communication. Wynne also knew that it increased costs that he also hoped to contain. As the nominal leader of the Bellevue-Yorkville Demonstration Project he increasingly steered its focus to the best and most efficient administrative practices in public health. Increasingly, Bellevue-Yorkville had moved to testing different forms of administrative structures, new clinical services, and ways to organize the necessary drives to ensure the immunization of children and the adoption of periodic medical exams for adults.[15] As Edward Devine, the new director of the Bellevue-Yorkville Demonstration, explained to an audience of community members in 1930, although called a demonstration, "we are not demonstrating anything. . . . We are carrying on as an experiment station."[16] Bellevue-Yorkville had moved beyond an emphasis on the wider spread application of known knowledge to a place that would generate new knowledge.

The work at the Bellevue-Yorkville Demonstration presaged many of the initiatives that would later be transferred to neighborhood health centers. It reorganized the Yorkville Tuberculosis Clinic into a Chest Clinic built around a new X-ray machine for diagnosis and a new system of records for follow-up visits. It forged links with the Bellevue Medical School Clinic that developed a new system of contact tracing for instances of venereal diseases. It established a Diagnostic Cardiac Clinic for children in an attempt to identify and treat those cases of rheumatic fever that they believed to be a leading contributor to the place of cardiac disease as one of the leading causes of adult mortality.[17] By 1932, the Bellevue-Yorkville Demonstration boasted of other leading accomplishments now increasingly practiced by the Department of Health throughout the city.[18] This included the department's first generalized nursing service that was to be the cornerstone of the health center movement. Indeed, the "initiative and the perseverance of the Bellevue-Yorkville nurses," its 1930 report noted, "was the most important factor in the successful operation of the whole service." Bellevue-Yorkville, in fact, had enlarged the scope of generalized nursing to include recreational as well as social, mental health, and nutrition support to families. "In public health work," it concluded, "the ability of the nurse to judge the problems of a family as a whole rather than just deal with one factor in the situation is an unquestioned advantage."[19]

The private Welfare Council weighed in with its own opinion. Its 1929 report, *A Health Inventory of New York City*, presaged the changing healthcare landscape. Constructed by the well-known healthcare reformer, Michael Davis, the inventory noted the problems that spurred the development of health demonstration projects: the lack of public and private coordination; the inequities of the distribution of health services that saw Manhattan with only 30 percent of the population of New York City served by two-thirds of all the private agencies; and services developed with little reference to a neighborhood's needs.[20] But he also noted changes that he believed to bode well for the future. He was impressed with the rise in the number of hospitals whose own outpatient clinics took health prevention and care coordination more seriously. He also saw the sharp increase in numbers of individuals across the city using these clinics.

And he believed in what he called the "dissolving" boundaries between private medical practice and public health promotion as individual physicians slowly incorporated medical exams and health teaching into their adult and pediatric practices.[21] Davis was less enthusiastic than many about the plans to carve the city into health districts. The entrance of hospitals as increasingly important institutions in the healthcare area, he believed, had a "radical" effect on the delivery of healthcare services and diminished the need to think about a health center in those areas well served by these institutions. Perhaps, he speculated, it might be better to think about health districts only in relationship to the need for home visiting nursing services.[22] This, of course, echoed the structure of the Henry Street Settlement and Visiting Nurse Service with its branch offices throughout the borough of Manhattan.

This recommendation also fit well with the direction and ambition of the East Harlem Nursing Service advisors, many of whom, such as Bailey Burritt, Homer Folks, Hazel Corbin, Lillian Wald, and Amelia Grant, also served as consultants to the *Inventory*. But Burritt, in his role as director of the AICP, also forged a link with the health clinics his association supported, not only in Columbus Hill but also in other underserved areas of the city such as Bowling Green and Mulberry Bend. The AICP, he wrote now Mayor Jimmy Walker, could provide the city with a "tested plan," not an "experiment" in how to organize and implement its health centers.[23] The first neighborhood health center opened in Harlem in 1931 to serve what had been a largely neglected and increasingly disaffected black population suffering from high rates of tuberculosis and maternal and infant mortality. Some monies had been set aside before the Depression for the construction of several additional health centers in "so called sore spots" in Manhattan on the lower West Side, in Mott Haven in the Bronx, and, in Brooklyn, in the combined neighborhoods of Red Hook and Gowanus and Williamsburg and Greenpoint.[24]

But LaGuardia's success in garnering federal dollars breathed new life into this movement. It also breathed new life into the Rockefeller Foundation's long-standing wish to more closely align government-funded public health departments with public health education in medicine and nursing. There was precedent: The two leading schools of public health, at Harvard and Johns Hopkins, had affiliations with local departments of health in nearby neighborhoods. But there was also a history. Wynne had four years earlier approached New York University about a possible affiliation and found "absolute opposition" from the University because of the politics surrounding "unpredictable" relationships with the city's Health Department, then staggering under allegations of widespread graft and corruption.[25]

But by 1934, Cornell University emerged as a possibility. A Rockefeller-funded pathologist, Eugene Lindsay Opie, wanted to extend his research on tuberculosis to the neighborhood surrounding Bellevue-Yorkville,[26] and Cornell itself hoped to develop an undergraduate department of public health. Although "loath to put itself in the hands of the city's Health Department," the University did agree to begin negotiations if it could appoint a "strong" professor of public health to the Health Center and could create "satisfactory" teaching arrangements for its public health and medical students. By 1935, the Foundation felt confident enough in the eventual success of these negotiations that it reserved $240,000 for the eventual creation of the "Cornell University Medical College Health Center of the New York City Department of Health." The Milbank Memorial Fund agreed to support this proposal by donating the monies it received from the sale of the then Bellevue-Yorkville Demonstration building. All participants joined in believing in the potential of this Health Center to become an "exceptional" urban health center and teaching site.[27] The Rockefeller Foundation also hoped it would solve the problem that the East Harlem Nursing and Health Service had become for it.

Public Health Nursing in the City

In 1934, New York City's Department of Health commissioned a study on "Some Special Health Problems of Italians in New York City" in conjunction with two newly planned, federally funded neighborhood health centers uptown in East Harlem, and in Mulberry Bend, an area west from the Bowery to the Hudson that at the time was the neighborhood with the second largest concentration of Italian and Italian American residents in the city's Lower East Side. By all mortality measures, the residents of East Harlem experienced "distinctly better" rates than those in Mulberry Bend. The overall mortality rate in East Harlem was 10.86 per 100,000 versus 12.84 in the city as a whole, a "most creditable" achievement. Residents of East Harlem died from pneumonia, diabetes, cancer, tuberculosis, influenza, and communicable diseases. But the rates of death from these diseases remained lower than in the city as a whole. Only diabetic deaths remained higher but, as the report pointed out, these rates still remained lower than those in Mulberry Bend. Mothers, infants, and children also died at rates lower than that of the city. The reason seemed apparent. It was because of "the intensive health work carried on in the district by the East Harlem Health Center." The shadow of that past project extended further. "There is every reason to believe," the report concluded, that the new downtown health center would "improve health conditions in that district to a considerable extent." No longer would its residents die needless deaths from pneumonia, tuberculosis, venereal diseases, and diabetes.[28]

Of course, the driving force behind this success lay with the work of its public health nurses, often acknowledged in print reports and memorandums but, as in the past, rarely refracted in data. They provided the bedside care of individuals with pneumonia, visited pregnant mothers and their infants in their homes, cared for individuals with tuberculosis and taught their families how to prevent cross-infections. By 1934, 75 percent of all the care and education delivered by the nurses at the East Harlem Nursing and Health Service were to mothers and their families in their own homes. They had incorporated communicable diseases into their traditional tuberculosis practices. They worked closely with physicians to implement plans for periodic medical exams for men, well-baby checkups for children, and prenatal exams for mothers; but their own role was to work with individuals and convince them to access follow-up care if "defects" were found. Examinations without such follow-ups were "futile" and nurses needed the "time and ability and energy" to make this happen for children as well as adults.[29]

East Harlem, still relatively financially secure, staved off the immediate effects of the Depression. Bellevue-Yorkville, dependent on nurses whose salaries were paid by the city's Department of Health, could not. As early as 1932, it had reduced the hours of its mental hygiene consultant and began practicing "rigid economies." Bellevue-Yorkville had also begun an "experiment," sending letters home requesting that mothers come to its schools to discuss "defects" found in their children's health exams rather than, as in the past, going to families in their own homes. By 1933, the Demonstration's experiment had shown what they framed as "most encouraging results." Thirty-seven percent of mothers who received such a letter took advantage of such appointments and came more prepared to address their children's health needs. As the convention in both private and public health nursing recorded a "visit" as a nurse calling on the family at home irrespective of whether a parent was home or answered the door, this report remained silent on whether or not this strategy reached more or fewer families. But one brief survey that tried to capture points of resistance to school visits found that 29 percent of families reported that they had never received a letter from their child's school; 28 percent said they were "too busy"; 15 percent had to nurse a sick family member; and in 9 percent of families both parents worked. Parents had again structured their own, often multiple places of action, some incorporating active points of resistance to outsiders inserting themselves into private matters and others reflecting the realities of their busy and often over-burdened lives.

School, rather than home visits, however, had as important an impact on public health nursing practice. Ostensibly much remained the same. When parents did come to the schools, the nurses used such opportunities to talk with mothers, not only about a particular child's health, but also about the physical, social, and economic difficulties the family as a whole faced. The home visit still had a place, but it was a much shorter one reserved for instances when nurses suspected children of experiencing infectious diseases or displaying difficult behavior problems. Those involved recognized that valuable information was lost when nurses did not visit a child's home. But the Bellevue-Yorkville nurses did find parents' attitudes "more satisfactory" when they voluntarily came to the schools.[30] The convergence of the Depression and Leonard Covello's wish to push health education out of systems of social welfare and charity did create more choices for parents reluctant to invite strangers into intimate family spaces. But this new system pushed to the margins those most difficult to reach and those most in need of assistance. Tensions had long existed between the city's public health nurses who had to—by law—deal with all "troublesome cases" and those at private agencies—such as those at East Harlem—who had more freedom to choose the individuals and families with which to work.[31] But now school rather than home visits gave the city's public health nurses more control over their practice. Moving forward, the city's school nurses now had a mechanism that—if they so chose—could keep at bay those they were reluctant to engage.

Nursing East Harlem

With more financial stability, East Harlem tried to maintain its home-based focus. But the home and the family inside were changing. Because of immigration restrictions, by the mid-1930s 60 percent of the population of the East Harlem Nursing district was born in the United States and only 30 percent had been born abroad; and the Service had noticed a decreasing demand for Italian translators.[32] Birth rates to young parents had plummeted more than 50 percent; and families were growing smaller in size, a trend abetted by a neighborhood birth control clinic and noted with approval as the East Harlem nurses felt that the children received more and better attention. Infant mortality had fallen to 56 per 10,000, as compared with 74 in 1923. But maternal mortality remained more intractable: its prevalence remained the same as in 1923.

As significantly, almost overnight hospitals had replaced homes as the preferred site of births and physicians had replaced midwives as the preferred attendant. Up until 1927, 85 percent of births had been in the home; by 1934, 65 percent occurred in hospitals. "Young mothers," Anderson reported to her board of directors in 1935, "look upon hospital care quite differently than did their foreign-born parents." Two outpatient medical clinics (including one nested inside East Harlem's Nursing and Health Service, staffed by nurses from the Maternity Center Association) had closed by 1933 as physicians' care now came though hospitals, hospital-based outpatient clinics, or private medical practices. The East Harlem nurses' first responsibility now was to ensure that families registered at the hospital in which they hoped to deliver as soon in their pregnancy as possible. Then they would begin their own work. Their overarching goal remained a safe pregnancy for both mother and child. But their "new approach" also meant using the prenatal and postpartum periods of home visits as a "starting point for the continuous program of parent education and child health supervision."[33]

Declining birth rates meant fewer children, but, Anderson continued, more intense involvement with families. Each family, on average, received nine visits during the prenatal and postpartum period from an East Harlem nurse: Blood pressure readings and urinalyses were also part of the visits. The overall average number of visits for all families combined approached twelve, suggesting some families received more intense scrutiny than others. Mothers who, according to the nurses' own, personalized evaluations, seemed "most alert" to the nurses' message received additional support in group meetings at the Service; those deemed "less alert" came to the Service for additional, individualized meetings. Fathers could participate in a "Fathers' Club" where a male leader led discussions about marital relationships and family problems. The Service also started a new Child Health Conference where mothers gathered with all their infants and preschool children to learn about normal growth and development, nutrition needs, the need for appropriate recreational activities, and to socialize with each other. All told, the Service boasted about reaching three-quarters of the districts' childbearing families.[34]

Anderson publicly boasted that East Harlem successes with families rested with her nurses' "techniques and sympathetic approach" that successfully "draw the parents out, recognize and give meaning to their experiences, direct them to knowledge or agencies where help may be secured, and yet leave them with a feeling of freedom in making their own plan."[35] But some other nurses at the Nursing Service refused to incorporate these tenets into their practices. Like their physician colleagues when approached to incorporate periodic medical exams into their practice, they had not been trained in the precepts of mental hygiene in their training schools and saw no need to learn them now. Sybil Pease, still East Harlem's consultant in mental hygiene, spent a considerable amount of her time working with these nurses. Anderson, in fact, believed

Pease's "best work" was with "those who needed considerable personal help to be intelligent workers in the mental hygiene field."[36] Some nurses, like the families they served, drew limits around the psychological intrusiveness required to practice this "new approach."

Nor was this new approach as easy to practice as it was to learn. As Anderson fretted to Mary Beard, still a staunch Rockefeller Foundation supporter, in 1937, "nurses are not born teachers." In fact, she continued, "for the most part they are particularly poor teachers until they have the time and opportunity to learn what and how to teach." Her challenge was to help her nurses incorporate the Service's "slogan" of "helping parents help themselves—help them formulate their own problems, ask their own questions, and then see how busy they can keep us trying to answer them."[37] Anderson constantly worried about the place of mental hygiene in public health nursing, in general, and the Service's responsibility to the mental hygiene movement, in particular.[38]

This emphasis on helping parents help themselves, in fact, slowly turned the Service away from its tradition of home visits toward clinic-based ones: Families who wanted to help themselves would take the initiative to seek out the help offered by the Service. By 1936, 41 percent of all services offered took place in clinics and health conferences inside the Service. "We aim to eliminate," Anderson announced to her board, "as far as possible, over-visitation in the homes and to encourage more activity on the part of families themselves" to come to the Service for classes.[39] Financial exigencies had now fully merged with middle-class expectations of initiative and independence. Beard noted to the Rockefeller Foundation on her routine visits to the Service "that the whole house was filled with the activities." Indeed, she continued, "all mothers were there by appointment, the appointment system having been as thoroughly developed here as in any private doctor's practice."[40] Yet while Anderson dreamed of a future in which the Service could develop "the methods of basic nursing services, of progressive education, and of case work," parents dreamed of one that included more recreational activities, like dances, and of turning the roof of the building into a "play school" for their children.[41] They dreamed of one geared more to their social than their health needs. Boundaries between health and welfare existed for clinicians, not for the families they served.

Nationally, critiques were developing as leading public health nurses in agencies across the country adopted this "new approach" that increasingly focused on the interior psychological life of their parents. Some public health nursing leaders, while supportive in concept, worried that this emphasis on the science of mental hygiene left nursing as vulnerable as medicine to charges that it had lost its sense of social justice: that the turn away from external health threatening environments blinded nurses to the real causes of health and ill-ness.[42] Others worried that the drift away from traditional bedside nursing and toward teaching in families' homes weakened nurses' place in the public health hierarchy. Thomas Parren, the surgeon general of the United States, and Mary Roberts, the editor of the *American Journal of Nursing*, joined ranks in 1939 by urging public health nurses to "get back to the middle of the road" by reuniting the more "concrete and the educational" functions of nursing. Roberts went one step further: She encouraged public health nurses to return to hospitals for postgraduate courses "in the newer methods of caring for patients."[43] Donald Armstrong, late of the Framingham Study and now the vice president of the Metropolitan Life Insurance Company, was more direct. Physicians, he warned public health nurses, do not understand the notion of the "talking nurse."[44]

The Changing Landscape

But these were not the only families who needed the Service. More and more, families from Puerto Rico had settled in East Harlem in the same old-law tene-ments that a second generation of more middle-class Italian Americans had fled by moving to the developing Bronx. And, in the midst of a Depression these families experienced the same lack of access to good jobs that would feed and clothe their children. This dismal environment, not surprisingly, brought with it a resurgence of tuberculosis, other infectious diseases, and shockingly high rates of maternal and infant mortality in East Harlem.[45] Many, if not most, of these families settled in the southern section of East Harlem, the neighborhood the Service had suggested access the closer Department of Health clinic when it moved to its new Kips Bay building. But the experiences of nurses in Colum-bus Hill also suggested not only a certain weariness in the face of a resurgence of an old enemy, but also a wish to avoid, again, the complicated intersec-tion of race and class that came with changing neighborhoods. Columbus Hill had undergone its own demographic transition as black families from the West Indies had moved out and those from Virginia, Philadelphia, and Newark had moved in. In the eyes of these nurses, these newcomers were more akin to a "cabin type negro" who had "no idea of housekeeping, home making, hygiene, or their privileges as citizens." There was "no question," they continued, that public health nurses were the best experts to build both "healthy men and women" and "conscientious citizens." But they felt they were working against "great odds." Though their staffing remained constant, their caseloads were decreasing because these families needed so much of their time. They wanted to expand the boundaries of their district to include the neighborhoods to which their West Indies families had fled. These older families, they insisted, were just at the point "where intensive health education can really begin." These families, they concluded, were "well worth their time and attention."[46] Whether at Bellevue-Yorkville, East Harlem, or Columbus Hill, public health nurses unwittingly withdrew resources from families that would not follow their advice to concentrate on those who showed more interest in their physical and mental health messages.

Traditional sick nursing care seemed increasingly irrelevant to their work.[47] It was not that there were no calls for such services. In 1936, for example, East Harlem received 1,070 such calls. But now the vast majority were for children with respiratory illnesses who had fewer hospital resources rather than adults did.[48] And hospitals were an increasing part of the healthcare landscape. These included the large, teaching hospitals famous throughout the country. But more worrisome were the small, private maternity hospitals sprouting up where mothers went to give birth—and, often, to die. Rates of maternal mortality in the city remained stubbornly high—higher, the Maternity Center Association (MCA) constantly pointed out, than any other civilized country. In Bellevue-Yorkville, they remained at approximately 6.2 deaths per 1,000 from 1922 to 1929, particularly problematic since mothers in the neighborhood's MCA clinic had experienced only 2.4 deaths per 1,000.[49] Similarly, they were 4 deaths per 1,000 in East Harlem where MCA also had a clinic. Yet in the city overall there were 5.1 deaths per 1,000.[50] Public health officials knew that these broad numbers needed some "nuance": in some parts of the city maternal mortality was a "negligible" concern.[51] But, they believed, in poorer neighborhoods maternal mortality was a "heart-rendering problem."[52]

And they believed they knew the cause. As both George Kosmak, the chair of MCA's board of directors, and Lindsey Williams, the new commissioner of health for New York City, constantly reiterated as they joined to launch a new and national educational campaign in 1930: A hospital confinement is not necessarily a safe confinement.[53] But it took the release of a 1934 study by the prestigious New York Academy of Medicine to drive that message home. The Academy surveyed the causes of all instances of maternal mortality in 1930 and 1931. It placed responsibility for two-thirds of all causes of maternal mortality at the feet of "incompetent practitioners" and on the fact that "conditions in hospitals [were] far from what they should be." It reported the dramatic increase in "operative deliveries, especially caesarian sections," and anesthesia by physicians with little experience. It noted the lack of any public or medical oversight on issues other than "minor points" of sanitation. It did not let mothers off the hook: 36 percent of maternal deaths were believed to be caused by their reluctance to seek prenatal care or their turn to induced abortions. Midwives still attended 10 percent of the births in New York City and their "meager supervision" resulted in the final 2 percent of all causes of maternal mortality.

Under MCA's leadership, a day-long symposium on "Community Responsibility for Improving Maternity Care in New York City" gathered together leading physicians, nurses, reformers, and interested lay women in 1934. The reports of the afternoon roundtable discussions revealed the intransigency of the issue of maternal mortality among poor women. A eugenics argument was brought to bear: The high rates of Caesarean sections could be explained by the "problems of certain kinds of men and women mating." An argument about a flawed study design was proposed: The criteria used to judge whether or not a maternal death could have been avoided was "utopian" and did not account for women arriving in an already "hopeless state" to municipal hospitals already overcrowded and inadequately funded. A gentler economic argument emerged: Poor women could simply not afford the thirty to fifty hours of pre- and postnatal care that the MCA deemed adequate. Mary Beard advocated for well-trained and supervised lay midwives, citing Scandinavia's example. But Alta Dines, speaking in her role as head of the Bureau of Nursing for the AICP, noted that public health nurses were in the perfect position to educate poor mothers in their homes about the importance of pre- and postnatal care. But, this kind of personalized care was always the first to be cut when agencies confronted tighter budgets.

Dines also took her discipline to task and echoed what was increasingly appearing in the nursing literature. Most nurses, she pointed out, did not like obstetrical nursing work as it was unpredictable, time-consuming, and labor-intensive. More damning was a 1931 brief report on the White House Conference on Child Health and Protection's Subcommittee on the Education of Nurses, which warned that there is "no escape from the conclusion that nurses do not know what adequate maternity care is."[54] And even the discipline's own *American Journal of Nursing* bluntly stated in a 1933 editorial that "mothers are dying because sick nurses are not taking proper precautions."[55] In the end, Dines concluded, "the nurse cannot be eliminated from taking her share in poor quality work."[56]

East Harlem's nurses did have additional maternity training even as it moved its program out of homes and into clinics. While its neighborhood maternal mortality statistics remained strong, its principled rationale for clinic-based work also had a more pragmatic base. Later, Grace Anderson would describe the years between 1933 and 1936 as a period of "retrenchment and consolidation." Even though the Rockefeller Foundation support remained constant, the average yearly income of the center had declined 23 percent. Some of this decrease came from an inability to collect sliding-scale fees charged to patients in a time of great need. But the Milbank Memorial Fund also dropped its contribution to the teaching service by one-third in 1933, by another 12.5 percent in 1935; before raising it again to the 1933 level in 1936. For the first time, the Service found itself with a $5,000 deficit as it closed the 1935 fiscal year. In response, Anderson had to cut both nursing staff and staff salaries, consolidate the nutrition and parent education consultants' role into one position, and place Sybil Pease on part-time employment.[57]

Anderson also terminated the role of the Service's statistician to cover her budget gap. This was a large loss. Rather than research new problems such as reaching out to families most in need, the Service's public health nurses now published pamphlets more akin to policy and procedure manuals than hard data. These pamphlets were popular. Those such as the *East Harlem Health Workers Handbook on Infant Development, Care, and Training* (1932) or *What Every Family Health Worker Should Know* (1934) or the *Handbook on Child Care* (1937) provided public health nurses across the country with the physical and psychological assessment data; with the forms used to collect and order data; and with the pamphlets left with families for their continued education. These were a valuable and valued service to the discipline of public health nursing. But it turned the Service away from its mission of research.

These decisions were made to save the teaching service. It remained absolutely intact, continuing to serve approximately one hundred full-time, part-time, and visiting students throughout the years of "retrenchment and consolidation."[58] But a larger problem loomed. Rockefeller Foundation funding would end in 1936 and, despite concerted attempts to explore future sources of income from federal sources, including the new Social Security Act and other New Deal programs that supported public health nursing education, no alternatives presented themselves.[59] The fight to save the Service depended on getting the Foundation to change its mind.

The Fight to Save East Harlem

The fight to save the East Harlem Nursing Service, in general, and its teaching mission, in particular, fell to Mary Beard, who, after the reorganization, was the Foundation's associate director of the International Health Division, now the only division within the Foundation that had any interest in nursing. The IHD had been continuously supporting women who might assume leadership positions in the countries where the Foundation had made an investment in medical sciences and public health by awarding fellowships to study in the United States. The East Harlem Nursing Service, along with Vanderbilt's, Yale's, and the University of Toronto's School of Nursing, had been consistently part of the fellowship experiences. The support of the IHD was critical.

First, Beard called a dinner meeting of the board of directors of the East Harlem Nursing and Health Service at the women-only Cosmopolitan Club on March 13, 1935. The discussion stretched past midnight, but the strategy for approaching Frederick Russell, the head of the IHD who had succeeded Pearce after his 1930 retirement, was finalized. Granted, Beard wrote Russell two days later, the Foundation had no interest in graduate education for public health nursing leaders, but sustaining the work of the East Harlem Nursing and Health Service was "one of those decisions which sometimes have to be made which are exceptions to the rules." It stood "head and shoulders" above any other teaching center. It was a far more superior developed practice field for public health nursing than, she emphasized, any of the schools of nursing in which the Foundation was interested, including the University of Toronto, a Foundation nursing favorite. It broke new ground in working to prevent mental illnesses through its family health teaching; and had "become the very kind of practice field which the IHD is attempting to foster all over the world."[60]

"Public health nurses," she continued, "cannot be educated without such a teaching field." But, keenly aware of the Foundation's aversion to fund any project that lacked independent sustainability, she tried to broker a compromise.[61] At present, she argued, there were no schools of nursing in the city sufficiently independent of hospital or medical school control that could absorb the graduate program at East Harlem. But in five years, she predicted, there would be. Right now, Teachers College and East Harlem represented the only counterweights to traditionally structured schools of nursing. But in five years and with an additional $90,000 grant from the Foundation, East Harlem could join the ever-strengthening Presbyterian and the New York Hospital Schools in their quest to offer postgraduate public health nursing education.[62]

Russell was, in fact, sympathetic to Beard's appeal and aware of the importance of the teaching service.[63] But Appleget, the Foundation's vice president, was less so. The Foundation officers debated the merits of all possibilities, including an affiliation with a proposed School of Public Health Nursing at Cornell, but such seemed "ambitious and complicated" and, as Cornell was now affiliated with the New York Hospital that had its own School of Nursing, not an option for the foreseeable future. On June 24, 1935, after repeated queries from Folks about the length of the deliberation, Appleget informed him that there would be no additional Foundation support. The Foundation would stay with its tradition and with the time-limited appropriation promised in 1932. The last grant of $10,000 would begin, as planned, on December 1, 1935

and end on November 30, 1936. He found himself unable to make any exceptions, no matter how worthy.[64]

Beard found it "most distressing" that this decision would leave the Service with only the pledged income from the four organizations that provided
financial or in-kind nursing resources to it—certainly more than half of its budget, but leaving initiatives in mental health, nutrition, and parental education
"crippled."[65] In December, she again approached the Foundation, suggesting
$5,000 to maintain these services in 1937 and 1938 until a plan could be devel-
oped that would transfer the East Harlem teaching staff to a new city health
center run by the New York Hospital–Cornell Medical Center.[66] The Foundation agreed to an additional year.[67] Beard then resigned from its board. "I feel I
can serve East Harlem better," she wrote to Folks in January 1936, "by resigning
from the Board than if I continue to be a member of it."[68]

The directors and staff of East Harlem refused to see this as the end of their
grand experiment. By 1937, they had prepared extensive materials to publi-
cize their work. The *East Harlem Nursing and Health Service: Fifteen Years of
Cooperative Endeavor: Should It Go On?* carefully laid out the steps taken to
achieve its "new approach to health work." Over the past fifteen years the Service had met the needs of the community for sick and maternity nursing. It had
developed new services such as health work for preschool children. It experi-
mented with the organization of nursing services. It had integrated knowledge
from nutrition and mental hygiene into all aspects of health work. And it had
brought the skills of a family caseworker to bear on common problems and
trained a new generation of practitioners from across the globe. They had bat-
tled what they saw as the illiteracy, old-world customs, and fatalistic indiffer-
ence of southern Italian immigrants, and now pointed with pride to how the
now-adult children they served brought their own children to the Service. They
felt they confronted what they believed other agencies knew but never publicly
admitted: the often ignored fact that the families most in need of service were
often those least likely to benefit from it and now more "consciously" selected
parents most responsive to teaching and guidance. While they continued to
attend to all families who experienced episodes of illness or the birth of a new
child, "maximum help" was given to families of "more ability."

They also believed that there was now complete acceptance by both the
public and other public health disciplines of the nurse as a "general practitioner" maintaining high standards of work that integrated the specialized services
of sickness nursing, maternal and infant nursing, and tuberculosis nursing. The
path had not been an easy one. Nor had they yet to claim complete success. The
pull toward specialized knowledge in such particular areas as obstetrics, child

development, and infectious diseases had created awkward language both at East Harlem, in particular, and in public health nursing, in general, about the need to "generalize about a specialty" or to create a "modified generalist" in which one nurse might claim specialized knowledge even as she met all the needs of her neighborhood.[69] But the nurses at East Harlem claimed they had elided such problems. Through constant experimentation, a "new approach" to family health work had been achieved, they believed, that integrated the work of the visiting nurses of the seriously ill and new mothers with the approach of the "health" nurse that—using new knowledge from nutrition and mental hygiene—would continue to reduce the need for sickness care. The East Harlem Nursing Service drew a sharp distinction between its "new" work and that of a previous generation of public health nurses. Its work was based on the individual needs of individual families, not on initiatives that would affect the community as a whole.

And the data they presented indicated it worked. Over the fifteen years the Service worked with families, malnutrition in children declined from 26 percent of all children it saw to less than 4 percent in 1936. Infant mortality declined from seventy-one deaths in 1923 to fifty-six in 1935. With its successful immunization initiatives, diphtheria deaths had almost disappeared and the measles death rate had significantly declined. The Service added careful caveats to this data. Many other agencies, it acknowledged, worked in the district, and the constant and consistent availability of federal relief dollars put more food on families' tables than the inconsistent earnings of wage workers before 1929. Still, in their minds, the best data were less tangible. It took pride in the changed relationships that existed between its nurses and their families; in parents' increasing ability to work through many problems on their own; and on the Service's prominence as a "laboratory" for the training of public health nurses.[70]

There was still much to be done. They "deplored" the fact that a move into administration was the only way nurses could increase their salaries, and wanted to create a new "senior field worker" so that their best and most experienced nurses could remain in "direct family health service." They wanted their student service "relieved" of its "most serious handicap—the pressure of bedside nursing." Under the terms of its agreement with Teachers College, the students "cannot carry the acute work," and, with the commitment of the Service to generalized nursing, "the burden of this falls on the advisors."[71] They also wanted to strengthen its mission of experimentation and publication. And their colleagues in the wider public health nursing world agreed. All of the letters of support it marshaled in support of its continued existence spoke to this unique function of the Service. Katherine Tucker, now the director of the Department of

Nursing Education at the University of Pennsylvania, wanted the Service to turn to studies of school nursing—an area of practice in which she believed there was little evidence based on real study and experimentation.[72] Marguerite Wales, now a consultant in nursing education to the W. K. Kellogg Foundation, noted that nowhere else have specific problems in public health benefited from the group thinking of experts, not just thinking about but actually working to solve problems; and its publications were avidly read by nurses across the globe.[73]

But, ultimately, the fight failed. As Appleget reminded Folks in his final appeal for continued funding in November 1937, the IHD only worked with sustainable governmental agencies, not voluntary ones like East Harlem. Its uniqueness, in fact, was its problem. It was neither a city health service nor an affiliated unit of the "great medical centers." It was providing a "notable community service," but that which made it renowned also made it vulnerable. East Harlem had been, in his mind, "rather stubborn in its independence." And it had been. It had kept itself free from relationships with hospital-based schools of nursing whose inevitable and insatiable demands for service would have compromised its ability to identify, experiment with, and solve what it saw as problems uniquely within the domain of public health nursing. And it had steered clear of the politics of public health by not seeking a relationship with the city's Department of Health. As Katherine Tucker pointed out in her letter of support, it had fewer "entangling alliances" and never suffered from "the periodic upheavals that usually occur in most community services."[74] This, Appleget acknowledged, led to the excellent work of the Service—and to the question of survivability once the Foundation stopped the last remnants of its support in 1937.

Privately, criticisms of the East Harlem Nursing and Health Service emerged in public health nursing circles as word of the Rockefeller Foundation decision spread. In 1937, Katherine Faville, a Vassar training school graduate, the former dean of Wayne University in Detroit, and the new director of nursing at Henry Street's Visiting Nurse Service, wrote Anderson that planning for the future would remain problematic as it was never very influential in New York City itself. Raising money would be easier if it had data showing that it influenced practices at the city's Department of Health.[75] Nineteen thirty-eight was a very demoralizing year. Tensions between the Service and Maternity Center Association flared. Hazel Corbin wrote that she had carefully studied the Service's statistics over the past five years, and decided that the Service's work with pregnant mothers still left it "with a long way to go." The Service had failed to meet a goal of 75 percent of pregnant women engaged in prenatal care; and too many of these women used private hospitals with a "low standard of care." Anderson countered with an argument that many of the mothers for whom the Service cared "bitterly resent their pregnancies," and too often saw neighbors having children easily and without scientific care. The nurses at East Harlem lived close to the lives of their parents and children and "can't avoid seeing their problems." Corbin retorted that she "could not help but smile" at Anderson's response. In MCA's experience "they are all people and their hopes and desires and ambitions and fears are the same, regardless of what class one might care to put them into."[76] And even the pool of potential students demurred. Mabelle Welch, the associate director of the Service, admitted "a lack of energy in the field." Not even nurses in postgraduate public health programs, she conceded, wanted the kind of advanced fieldwork training that East Harlem offered. These students largely preferred only classroom lectures, and if their program did require fieldwork, they approached it as a "necessary evil."[77]

The East Harlem Nursing and Health Service limped along with increasing deficits for the next few years. Slowly, and inexorably, participating organizations began withdrawing. MCA pulled out. The Henry Street Settlement and Visiting Nurse Society, always an ambivalent partner, decided it needed to cut its appropriation; the Milbank Memorial Fund could not continue its financial support "indefinitely." No other foundations, including the New York and the Carnegie Foundations, could find their way free to provide support; and the Department of Health had too great a dependence on federal relief dollars to even consider a contribution. Slowly, the Service reduced its staff. Grace Anderson took an extended leave to deal with her "serious heart condition." And Maybelle Welsh, her assistant, began working part-time for Teachers College.[78]

The Service finally acceded to the inevitable. It gave all remaining staff an additional month's pay and closed. A personal and rather terse letter to the Foundation in January 1941, from Margaret Nourse, the president of Saint Timothy's League and long-time supporter of the Service, acknowledged that "your generosity and real interest in this project entitles you to know of the imminent shutting down of this teaching centre [sic]."[79] An innovative and interdisciplinary Nursing Service and, as Nourse inadvertently emphasized, teaching service, that had hoped to transform the practice and curricular landscape of public health nursing now shared the fate of the earlier Health Center and had come to an end.

Chapter 5

Not Enough to Be a Messenger

In the early 1920s, those who would reform the US healthcare system established a small series of demonstration projects that would provide patients access to high-quality, cost-effective, and expertly coordinated healthcare. Much like the goals of those supported by today's Center for Medicare and Medicaid Innovation, these demonstration projects would translate ideas into practices that could be easily adopted by existing healthcare structures. Many of the demonstration projects of the 1920s proved successful and many of the practices they supported proved enduring. They established the idea of a neighborhood health center as the most effective site of public health initiatives, an idea now institutionalized in the Affordable Care Act's expansion of monies for community health centers to bring high-quality care to poor urban neighborhoods and isolated rural ones.[1] They placed the practice of "periodic medical exams" for both children and adults as central to maintaining health and preventing illness. They set the stage for the eventual insurance coverage of exams such as mammography for women, prostate cancer screenings for men, and vision and hearing exams for children. They launched an emphasis on oral hygiene and dental care as key pillars in one's overall health status and, while availability of service remains less than ideal, an emerging field of research and practice in oral-systemic health holds the potential to alter this terrain.[2] These demonstrations, in fact, eventually established the current norms of primary care. And public health nurses played a critical role. They brought the messages of health into illness care even in the face of the often-suspicious communities they served. And they brought the norms of middle-class health practices to families excluded by the financial requirements of fee-for-service medicine.

More immediately, much good came from these demonstrations. The East Harlem Health Demonstration Project proved that its community sought health and welfare information when it was easily available to them. The East Harlem Nursing and Health Demonstration Project highlighted the importance of research in the public health nursing agenda. And the Bellevue-Yorkville Demonstration Project showed how public and private partnerships could be successful when carefully calibrated to meet the Department of Health's own mission and goals. Each provided different data streams that Shirley Wynne, New York City's commissioner of health, needed as he moved to implement a system of neighborhood health centers when federal construction dollars became available in the mid-1930s. Those involved in the demonstration projects constantly talked with one another and showed a certain nimbleness in stepping in to solve problems or provide resources that another lacked.

But notions of "coordination" and "cooperation" were, as contemporaries recognized, always easier to conceptualize than to implement. Michael Davis, the noted reformer, told nurses at the 1939 Annual Meeting of the National Organization for Public Health Nursing that these notions demanded "imagination and courage" rather than the "protective attitude" too often engendered by ideas for change. Despite massive federal involvement in providing health and welfare services during the Great Depression, there would still be a place, he continued, for smaller, private health agencies, like the East Harlem Nursing and Health Service. But he added an important caveat. To survive, he told public health nurses that they needed the "imagination to conceive, investigate, and define what unmet needs are" and the "courage to scrap past activities and work on that somewhat uncertain and often controversial borderline which runs between the present and the future."[3]

This borderline was an increasingly fraught space. Some of the tensions in this space, as historian Karen Buhler-Wilkerson has argued, did lie with a changing context. The closing of American borders to immigrants mitigated the need for a public health nurse to bring both "medicine and a message" of Americanization to poor families.[4] Indeed, the poorest families now crowding New York City—those from Puerto Rico and blacks from southern states—already claimed American citizenship. But these claims were tenuous, complicated, and, preferably, ignored. In East Harlem, in particular, Puerto Rican families did not fit comfortably into an established, entrenched, and binary racial hierarchy. While some of the neighborhood's politicians and activists, including Leonard Covello, found ways to form alliances with members of the Puerto Rican community, most others kept their distance. Puerto Ricans, in turn, kept their distance from blacks as a strategy to fend off further marginalization.[5]

This created a toxic neighborhood stew that erupted in the 1935 Harlem Riot. Its historian, Jeffry Stewart, described this as the "first modern race riot" in the United States and "symbolized that the optimism and hopefulness that had fueled the Harlem Renaissance had died."[6]

And the Great Depression only exacerbated the increasingly common turn toward hospitals for childbirth and surgeries, especially those for infected tonsils and adenoids and appendicitis. Hospitals also treated accidents (an increasing problem on the crowded streets of New York City), emergencies, and cases of pneumonia, digestive diseases, and the "degenerative diseases" of middle-aged adults.[7] Through the 1930s, the vast majority of these could still be managed in New York City homes, but the increasing depiction of urban apartments as small, crowded, and unsanitary created a pull toward the clean if not sterile environment of hospitals. Poorer families wanted what they saw middle-class families accessing and nurses encouraging, especially in childbirth. They, too, wanted healthcare moved out of their homes and into hospitals. And they needed this to happen, as well. Poor families, unable to afford private physician fees that would come with a home visit, turned to the city's municipal hospitals and outpatient clinics. Overcrowding in hospitals was rampant; Bellevue Hospital, for example, reportedly operated at 110 percent of its capacity by 1933. And long lines were common in their outpatient clinics, especially as the "Depression poor"—those whose own standard of living had been decimated by unemployment—joined others in seeking what was often free care. "People went to hospitals," historian Rosemary Stevens has argued about these institutions during the Great Depression, "expecting to be taken in."[8]

This chapter more deeply examines the policy implications we might learn not just from the demonstration projects themselves but also from the work of the nurses who were their public faces. There may be many lessons learned from the East Harlem and Bellevue-Yorkville Demonstration Projects in New York City—lessons such as the need for small, focused projects rather than "monumental" ones or the need for such projects to have carefully worked through arrangements with all the constituent stakeholders involved in the public's health. But by focusing on the possibilities and the problems that nurses confronted in their day-to-day work with families we see other lessons. In the end, the nurses in New York City's health demonstration projects did achieve significant successes. They, along with like-minded colleagues, opened public health nursing to interdisciplinary areas of knowledge long before such was popular. They introduced mental health concepts into the practice of nursing long before they became engrained in nursing school curricula. And they broadened their "new approach to health work" to be more inclusive of families rather than individuals. Yet their history also provides a cautionary message as we move forward to capitalize on the opportunities afforded by the Affordable Care Act (ACA) and the calls for proposals from the Center for Medicare and Medicaid Innovation. Disciplinary wishes—more specifically, the quest of public health nursing leadership for control over the education needed to enter their practice—cannot be separated from the needs of constituent communities.

Science and Social Justice

As historians Amy Fairchild, David Rosner, James Colgrove, Ronald Bayer, and Linda Fried have pointed out, the postwar turn to the "new science of public health" defined the field through the 1920s and 1930s. This new "science" took the laboratory's seemingly unbiased data and the individual as its domain. It joined with the profoundly conservative political and social climate that also produced the sharp immigration restrictions that characterized changes in the East Harlem neighborhood.[9] And, particularly at the Milbank Memorial Fund's three demonstration sites across New York State, the fear of being accused of being "socialists" or "radicals" by members of the American Medical Association significantly tamped their enthusiasm for doing what they believed were the right things the communities needed. They knew, for example, that lay public health officials often had the most prescient vision of what the demonstrations might do for their particular communities. But, they decided, they could not afford to antagonize local general practitioners and always chose physicians to lead more conservative initiatives.[10]

New York City's public health nurses were absolutely central to the success of this conservative and medicalized vision. They came to their support of this vision steeped in a training school tradition that had valorized medical science and medical knowledge not only as sources of truth but also as those of power and authority. Certainly, the site where they learned this knowledge—the hospitals that used their work to care for patients through their three-year diploma school experience—emphasized the kinds of knowledge needed for illness care. But it also shaped an enduring partnership in which nurses saw themselves as the "educated allies" of physicians and engaged in a more inclusive and more acceptable relationship with medical knowledge than that of women who would be physicians. White nurses would be assistants to powerful white men in ways that reaffirmed conventional gender and racial hierarchies.[11]

Nowhere was this more evident than in New York City public health nurses' embrace of the movement to ensure all mothers had medically supervised births in hospitals. Their almost embodied belief in the primacy of medicine and science took precedence over data about where and how mothers were dying. But they did good as well. Public health nurses were also in homes and on the street encouraging families to immunize and vaccinate their children. They translated the science supporting the Schick test for diphtheria, the Wasserman for syphilis, or the Mantoux for tuberculosis into language that individuals, and parents in particular, could understand. New York City's public health nurses, steeped in a training school experience that valued their command of medical knowledge, embraced this new science. They wove it into the "message" they delivered with their medicines to families.

Yet, the mantra of mental hygiene remained enigmatic to many of these scientifically trained public health nurses. As they surveyed the new knowledge available in the interwar years, they chose that which resonated most strongly with their training school experiences. Much like many physicians who refused to incorporate exhortations to include periodic medical exams into their practices because it lay outside their own training, many public health nurses in New York City chose to incorporate knowledge into their practices that reinforced the familiar. The incorporation of mental hygiene, later renamed mental health, would come to nursing. But it had to await the post–World War II movement to reorient nursing practice in ways that emphasized the primacy of the individual and the nurse-patient relationship.[12]

Interestingly, medicine was the one public health discipline largely absent from the day-to-day considerations of these nurses and social workers at both the East Harlem and the Bellevue-Yorkville demonstration projects. Nurses nursed and social workers created "adjustments" in individuals and families with little attention to the politics of practice raging above them. Physicians, in fact, seemed more concerned about the practices of nurses and social workers than the nurses and social workers were about medicine's—about nurses' need to scrupulously follow their medical instructions and about social workers' need to be wary about providing material relief to a family who would use it to purchase a new automobile for Sunday drives into the country.[13]

In many respects, public health nurses had already worked through their disciplinary tensions with physicians. By the early 1920s, both public and private public health nurses worked under sets of "standing orders" from physicians that covered procedures in most routine cases of bedside nursing and health teaching. Changes to these orders were incremental and, most often, controversial. Bellevue-Yorkville's plans for changing established nurse and physician relationships were the most ambitious. It offered private medical practitioners in its demonstration area their own public health nurse to follow patients into their own homes both to ensure they followed medical directives and to enhance messages of health. Most physicians refused this offer.[14]

But if these tensions had been resolved, those with newly emerging public health workers had not. Public health roles for nutritionists and health educators had emerged in the 1920s, but these roles, and the individuals in these roles, never seemed of concern to the city's public health nurses' deliberations: Nutritionists were too small in number; and health educators took groups in the community as their domain. Social workers who, as did nurses, took individual families as their community of interest, represented the greatest threat. But public health nurses, both in New York City and across the country, vanquished that threat with relative ease. They capitalized on the relative trust they had built with families and laid claim to their tradition as the only group of women health workers with legitimate claims to specialized medical knowledge.

But this victory had its own costs. Public health nurses joined others in turning toward science and away from what had been an earlier generation's robust sense of social justice. They were not alone. By 1934, Homer Folks found himself "disturbed" that the Milbank Memorial Fund's advisors were steering it into fields of "medical economics" and away from direct clinical initiatives in public health. He brooked no concessions when challenging the assertion that medical economics, particularly those concerning implementing some of the recommendations from the Fund-supported Committee on the Costs of Medical Care, were indeed related to public health.[15] And eventually even the venerable Henry Street Settlement and its Visiting Nurse Service had to separate into two autonomous organizations. By 1944, no one public health worker—whether she was a nurse or social worker—could maintain Lillian Wald's vision of health and social justice for communities in need.[16]

The experiences of public health nurses in New York City's health demonstrations do remind clinicians, in general, and nurses, in particular, that it is "not enough to be a messenger" of physical health and mental well-being. Decoupling messages of health from the material conditions that make health possible—from education, from housing, from gainful employment—creates a hollow message that, as other scholars have pointed out, inevitably blames victims for their ability or inability to make changes in their lives. But this story sharpens this message. Nurses were not immune to the effects of the complicated and intersecting domains of race, class, and gender. And within segregated race communities, class mattered most to both white and black public health nurses. It was not the only factor. In the 1920s, at the beginning of the health demonstration projects, both white Italian and black West Indian families could make legitimate assimilation claims on the health and social welfare agencies dedicated to both material relief and the process of Americanization. These claims gave these families the opportunity to accumulate resources

Figure 6. A Puerto Rican Family in the East Harlem Nursing and Health Service's Care

Reprinted with the permission of the Rockefeller Archive Center.

necessary to move away from their traditional urban neighborhoods into more suburban ones. Those families that followed them into the neighborhoods of the demonstration projects—those from Puerto Rico and the American South— occupied a more complicated social space with ambiguous assimilation claims, tenuous citizenship rights, and little access to the changing levers of political power. Moreover, they moved into a public health system that had severed the links between their health and their environments.

But both white and black nurses struggled to reach the Puerto Rican families in East Harlem and the southern ones in Columbus Hill. Their frame of assimilation and aspiration had disappeared. As significantly, both groups of middle-class women unwittingly reflected and refracted the nativist and racist assumptions pervasive in the conservative interwar years. They both believed their new constituents were unable to assimilate to, if not American standards, then to middle-class norms. This, too, is one legacy that New York City's public health nurses helped create in supporting a medicalized model of public health that incorporated prevailing social assumptions. But, with the full implementation of the Affordable Care Act—and especially as issues of access to prenatal care, poor maternal health outcomes, and efforts to reach preschool children remain problematic—we may be given another opportunity to recouple health with its social determinants for all in need.

Constituent Need and Disciplinary Interests

In many respects, the public health nursing leaders involved in New York City's health demonstration projects achieved all their disciplinary ambitions. They saw a 1923 Report of the Committee for the Study of Nursing Education enshrine their standards for nursing education first in the United States and then, a short while later, abroad. They celebrated the establishment of a separate Bureau of Nursing in the city's Health Department and the fact that, for the first time in the city's history, a nurse and not a physician supervised the practices of other public health nurses. They were among the pioneers of a "new approach to health work" that brought families into their disciplinary domain. Certainly, they never achieved the Rockefeller Foundation's goal of uniting public and private public health nursing agencies, but, in fact, these public health nurses never shared this agenda. They believed in the value of private agencies, like the Henry Street Settlement, to set the standards for quality and innovation that the city's own public health nursing bureau would soon follow. They considered that the Foundation had failed them in refusing to support East Harlem's postgraduate teaching mission rather than that they had failed the Foundation.

Yet, in the end, the East Harlem Nursing and Health Service's commitment to take practice and teaching as its explicit domains in 1928 held the seeds of its eventual failure. It may have met the needs of many of its patients, but it served the needs of a discipline looking to create well-educated public health nurses. The Service, in fact, lost its way when it became enamored with its teaching mission. Rather than performing research on new problems such as how to reach out to families most in need, it now published pamphlets more akin to policy and procedure manuals than hard data. These pamphlets provided public health nurses across the country with the physical and psychological assessment data collected by the Service's nurses; with the forms used to

collect and order data; and with the pamphlets left with families for their continued education. These were a valuable and valued service to the discipline of
public health nursing, but they reflected little innovation. Rather, they reflected
the practices of the more progressive Visiting Nurse Associations in New York,
Boston, Chicago, St. Louis, Toronto, and Baltimore. And they reflected little of
the changing healthcare landscape, including the increasingly prominent place
that cancer, heart disease, venereal disease, and chronic disease now had on the
public health agenda.

Indeed, the eyes of the nurses at East Harlem were on what they believed
their constituent families needed rather than on how they understood what
these families wanted. They decided to restrict their practice during the financial turmoil of the 1930s to only the more receptive families in their neighborhood and shut themselves off from others who, in all likelihood, may have
needed them the most. It closed itself to the voices of other constituents in its
community. And it reinforced the discipline's own insularity.

East Harlem did try to find alternative sources of funding for its practice and
teaching mission. At a 1934 meeting of the East Harlem Council of Social Agencies, Grace Anderson of the East Harlem Nursing and Health Service declared
that if the poor were to receive the help they needed, the city would have to
move beyond merely creating health centers. It needed to provide the same
subsidies to home nursing as it currently did to health centers and municipal
hospitals. These subsidies for "home relief," she argued, were as "legitimate
a charge to the taxpayer as hospitals."[17] Anderson was not alone in this wish;
she only echoed the hopes of leading public health nurses across the United
States. These kinds of subsidies never materialized. Rather, Anderson's hope of
municipal funding to preserve sick nursing and health promotion in the home
reflected healthcare as progressive public health nurses wanted it to be. They
wanted it to remain constructed within intimate personal relationships forged
in homes and not in the more impersonal ones found in the central hospitals
and healthcare centers that increasingly dominated the healthcare landscape.[18]

In the end, these demonstration projects have also left some unanswered
questions that we may now have the opportunity to address in the Center for
Medicare and Medicare Innovation's calls for its own demonstration projects.
The collapse of initiatives in the early 1930s to investigate the kind of worker
or team of workers to best deliver public health services at the point of contact with those in need has left fundamentally important ideas unexplored.
The disciplinary domains of nursing and social work—domains first forged
in hospitals—may not map cleanly onto the geographies of public health.
And the tensions and conflicts that existed between these disciplines may be

emblematic of struggles to assert dominance in a hierarchical public healthcare structure led by medicine. Or they may represent points of disconnect between what the disciplines wanted to do (and were prepared to do) and what their families needed. Certainly, we do see the rise of formal "health educators" in the 1930s, and the creation of roles for lay "community health workers" in the 1960s. But these newer public health roles remain layered upon a public health structure built around the joining of the disciplines of nursing and medicine that has had little sustained examination.

And the paradox of prevention remains. The ability to shift an entire community or population to behaviors widely acknowledged as healthier still remains highly problematic and contested. The day-to-day practices of public health nurses do lack the drama, the intensity, and the technology that sustain a community's interest. Yet, nurses in the community reached the community in ways that other disciplines could not and did not. The widely recognized and respected validity of their knowledge claims, in fact, situated nurses at the center of a matrix of competing public health agendas. The champions of a new public health science, the foundations that supported the demonstration projects, the families they served, and the other disciplines with which they worked all had ideas and projects that they believed nurses were particularly situated to implement. In some ways, the experiences of nurses in New York City's health demonstration projects suggest the paradox of prevention is as much about power as it is about policy. The experiences of nurses in New York City's demonstration projects suggest a process of constant negotiations around the ability to set, implement, limit, and financially support a health as well as well as an illness agenda.

As important, the nurses in New York City's demonstration projects added their own disciplinary agenda to this process. Their intent was to situate themselves at the nexus of independent, interdisciplinary, and instructional nursing practices. In many respects, these nurses were successful. Those at Bellevue-Yorkville achieved disciplinary independence when nurses, rather than physicians, obtained control of public health nursing practice; and those at the East Harlem Nursing and Health Service established an innovative teaching service that remained the envy of progressive public health nursing educators throughout the country. But East Harlem's critics were correct. Without alliances—however problematic—with either a school of nursing or the city's own Department of Health, East Harlem nurses had no formal power to press for health. Their independence had a steep price.

Closely examining the power of nurses to shape public health practices in the interwar years also calls attention to the influence of the less tangible goals, needs, and ambitions of the many different constituents that conceptualized, paid for, delivered, and received healthcare services. As the story of health demonstration projects in New York City illustrates, these goals, needs, and ambitions were as critically important drivers of ultimate success or failure as the theoretical underpinnings that led to their creation. These drivers—the tensions between public and private responsibility for setting public healthcare agendas; between lofty aspirations of coordinated care and the realities of not wanting to cede to a controlling authority; between the hopes of a discipline and the requirements of its community; between public healthcare as nurses wanted it to be and the healthcare landscape that actually existed—were as important in the health demonstration projects in New York City as were the clinical and economic metrics measured. The experiences of the East Harlem Nursing and Health Service stand as a seminal example. The Service gained international fame among public health leaders for its innovative and independent nursing practice and research. Yet it ultimately failed because its commitment was to a particular disciplinary mission that did not meet the needs of the constituent communities it served. From 1928 until its closing in 1941, the Service focused more on the educational advancement of public health nursing and less on addressing the real, changing healthcare needs of those in its East Harlem home. For all its successes, it also provides a cautionary tale as we consider the multifaceted dimensions of the clinical experiments that will be part of the Innovation Center's demonstrations in comprehensive, high-quality, and coordinated care.

Finally, the experiences of public health nurses as they sought to capitalize on changes in the public healthcare landscape in 1920s and 1930s New York City suggests that we need to have sustained debates about the educational and the regulatory frameworks that structure the practices of a "new kind of messenger." The issues are not that dissimilar. Public health practitioners still speak of interprofessional practices, community partnerships, the new epidemic of chronic diseases and the resurging one of infectious diseases. And the dialogue has already begun. In 2010, the *Lancet*, one of the most influential medical journals across the globe, had already called attention to the "social construction" of our current division of labor among public health professionals in its landmark call for transforming conventional educational structures for an increasingly interdependent world.[19] And the Institute of Medicine's *Future of Nursing* has most recently argued that nurses, in concert with other disciplines, need to reconceptualize their roles as health coaches and system innovators.[20] As the experiences of nurses in New York City's health demonstration projects illustrate, this dialogue cannot take place without a keen assessment of how the national healthcare landscape will change in the context of the Affordable Care Act and of how the global one will be transformed as the world becomes more interconnected and interdependent. Yet it will not be easy. However much we value the idea of high-quality, coordinated care, the history of health demonstration projects in New York City illustrates just how hard that can be when different disciplines, organizations, and associations have a vested interest in attending to their own advancement, place, and power as they legitimately search for better ways to care for the people in their care.

As the Center for Medicare and Medicaid Innovation continues to issue calls to test best practice models to increase access to high-quality, cost-effective, and coordinated healthcare, we should see them as an opportunity to reengage with the unanswered questions of these earlier demonstrations. This is the moment to consider what is core to the different public health disciplines and what can be shared with others. And this is the moment to remember that ideas engender change, but the prerogatives of gender, class, religion, and disciplinary interests shape their implementation. As nursing develops potentially exciting projects that can be "scaled up" to serve even more constituents, we might also remember that the processes and politics of practice remain critically important. The notion of "coordination" among different disciplines is very challenging to operationalize. But we now have another chance to do so.

Notes

Introduction

1. "Strong Start for Mothers and Babies Initiative: General Information," *CMS.gov*, last accessed August 9, 2015, http://innovation.cms.gov/initiatives/strong-start/.
2. Congressional Budget Office, "Lessons from Medicare's Demonstration Projects on Disease Management, Care Coordination, and Value Based Payments," *CBO.gov*, last accessed March 10, 2015. http://www.cbo.gov/ftpdocs/126xx/doc12663/01–18–12-MedicareDemoBrief.pdf.
3. For some examples, see James Colgrove, *State of Immunity: The Politics of Vaccination in Twentieth-Century America* (Berkeley: University of California Press and the Milbank Memorial Fund, 2006), 89; Daniel M. Fox, "The Significance of the Milbank Memorial Fund for Policy: An Assessment at Its Centennial," *Milbank Quarterly* 84, no. 1 (2006): 5–36; Elizabeth Toon, "Selling the Public on Health: The Commonwealth and Milbank Health Demonstrations and the Meaning of Community Health Education," in *Philanthropic Foundations: New Scholarship, New Possibilities*, ed. Ellen Lagemann (Bloomington: Indiana University Press, 1999), 119–130; and George Rosen, "The First Neighborhood Health Center Movement: Its Rise and Fall," *American Journal of Public Health* 61 (1971): 1620–1635.
4. Karen Buhler-Wilkerson, *No Place Like Home: Nursing and Home Care in the United States* (Baltimore, MD: Johns Hopkins University Press, 2003); Susan Reverby, *East Harlem Health Center: An Anthology of Pamphlets* (New York: Garland Publishing, 1985).
5. The Center for Consumer Information and Insurance Oversight, "Shining a Light on Health Care Insurance Rate Increases," *CMS.gov*, last accessed March 14, 2015. http://www.healthcare.gov/news/factsheets/increasing_access_.html.
6. Rima Apple, *Perfect Motherhood: Science and Childrearing in America* (New Brunswick, NJ: Rutgers University Press, 2006). There is a rich literature on this topic that also includes Apple's *Mothers and Motherhood: A Social History of Infant Feeding, 1890–1950* (Madison: University of Wisconsin Press, 1987); Molly Ladd Taylor's *Motherwork: Women, Child Welfare, and the State 1890–1930* (Urbana and Chicago: University of Illinois Press, 1994); Janet Golden's *A Social History of Wet Nursing: From Breast to Bottle* (Cambridge: Cambridge University Press, 1996); and Julie Grant's *Raising Baby by the Book: The Education of Modern Motherhood* (New Haven: Yale University Press, 1998).
7. Barbara Beatty, Emily Cahan, and Julie Grant, eds., *When Science Encounters the Child: Education, Parenting, and Child Welfare in Twentieth-Century America* (New York: Teachers College Press, 2006), 2.
8. Jodi Vandenberg-Daves, *Modern Motherhood: An American History* (New Brunswick, NJ: Rutgers University Press, 2014).
9. Jacqueline H. Wolfe, *Deliver Me from Pain: Anesthesia and Birth in America* (Baltimore, MD: Johns Hopkins University Press, 2009).
10. Kathleen W. Jones, *Taming the Troublesome Child: American Families, Child Guidance, and the Limits of Psychiatric Authority* (Cambridge, MA: Harvard University Press, 1999). The interwar years also saw public health initiatives using science

broaden its scope to include occupational health, venereal disease control, accidents and vehicular safety, cardiovascular disease, and cancer mortality. See John Wardo and Christopher Warren, eds., *Silent Victories: The History and Practice of Public Health in Twentieth-Century America* (Oxford and New York: Oxford University Press, 2007). For the move to incorporate oral health, see Alyssa Picard's *Making the American Mouth: Dentists and Public Health in the Twentieth Century* (New Brunswick, NJ: Rutgers University Press, 2009).

11. Andrew L. Morris, *The Limits of Voluntarism: Charity and Welfare from the New Deal through the Great Society* (Cambridge: Cambridge University Press, 2009).

12. Robert Kohler, *Foundations and Natural Scientists, 1900–1945* (Chicago: University of Chicago Press, 1981).

13. Ellen Lagemann, *Politics of Knowledge: The Carnegie Corporation, Philanthropy, and Public Policy* (Middletown, CT: Wesleyan University Press, 1989).

14. Amy Fairchild, David Rosner, James Colgrove, Ronald Bayer, and Linda Fried, "The Exodus of Public Health: What History Can Tell Us about the Future," *American Journal of Public Health* 100, no. 1 (2010): 54–63.

15. John Duffy, *The Sanitarians: A History of American Public Health* (Chicago: University of Illinois Press, 1992).

Chapter 1 — Medicine and a Message

1. See Karen Buhler-Wilkerson, *False Dawn: The Rise and Decline of Public Health Nursing in the United States, 1900–1930* (New York: Garland Press, 1989).

2. Patricia D'Antonio, *American Nursing: A History of Knowledge, Authority, and the Meaning of Work* (Baltimore, MD: Johns Hopkins University Press, 2010).

3. Municipal Archives, NYC DOH, H34.1, Roll 15, "Report of the Department of Health of New York for the Year 1919."

4. John Duffy, *The Sanitarians: A History of American Public Health* (Chicago: University of Illinois Press, 1992).

5. John Dill, "Who Shall Nurse the Sick?," *American Journal of Public Health* 11 no. 2 (1921): 108–112.

6. For a fuller discussion, see Susan Reverby, *Ordered to Care: The Dilemma of American Nursing, 1850–1945* (Cambridge: Cambridge University Press, 1987).

7. Buhler-Wilkerson, *False Dawn*. See also Diane Hamilton, "The Cost of Caring: The Metropolitan Life Insurance Company's Visiting Nurse Service, 1909–1953," *Bulletin of the History of Medicine* 63 (1987): 414–434.

8. Maternity Center Association (MCA), Box 56, Folder 1. For the debate in the public health nursing literature see, for example, Mary Beard, "The Attendant as an Assistant to the Public Health Nurse," *Public Health Nurse* 2, no. 3 (1919): 181–182.

9. In 1920, the need to sit for a state licensing exam was still voluntary in most of the United States, although that would begin changing by mid-decade. See Reverby, *Ordered to Care.*

10. D'Antonio, *American Nursing.*

11. For postgraduate programs, see Patricia D'Antonio, "Women, Nursing, and Baccalaureate Education in Twentieth-Century America," *Journal of Nursing Scholarship* 36, no. 4 (2004): 379–384. For content see Mary Sewall Gardner, *Public Health Nursing* (New York: Macmillan, 1919).

12. For patronage, see D'Antonio, *American Nursing*, 71.

13. Marjorie N. Feld, *Lillian Wald: A Biography* (Chapel Hill: University of North Carolina Press, 2008).

14. Kara Dixon Vuic, "Wartime Nursing and Power," in *Routledge Handbook on the Global History of Nursing*, ed. Patricia D'Antonio, Julie Fairman, and Jean Whelan (Oxford: Routledge Press, 2014), 22–34.

15. Cindy Gurney, "Annie Warburton Goodrich," in *American Nursing: A Biographical Dictionary*, ed. Vern Bullough, Olga Church, and Alice Stein (New York: Garland Publishing, 1988), 145–149.

16. Joellen Watson Hawkins, "Mary Adelaide Nutting," in *Dictionary of American Nursing Biography*, ed. Martin Kaufman (New York: Greenwood Press, 1988), 274–277.

17. Alice Howell Friedman, "S. Lillian Clayton," *Dictionary*, ed. Kaufman, 65–68.

18. Karen Buhler-Wilkerson, "Mary Beard," in *American Nursing: A Biographical Dictionary*, ed. Bullough, Church, and Stein, 19–22.

19. Rockefeller Archive Center (RAC), Conference on the Training of Nurses, Called by the Officers of the Rockefeller Foundation, February 28, 1920.

20. For a discussion of the Flexner Report, see Kenneth Ludmerer, *Learning to Heal: The Development of American Medical Education* (New York: Basic Books, 1985), 166–190.

21. Ira V. Hiscock, "The Development of Neighborhood Health Services in the United States," *Milbank Memorial Fund Quarterly* 13, no. 1 (1936): 30–51.

22. Community Service Society (CSS) #2073, Box 63, Folder: Home Hospital Planning, Confidential: The Future Home Hospital, October 18, 1921; also CSS #0273, Box 63, Folder: Home Hospital Planning.

23. Milbank Memorial Fund (MMF), Group 845, Series II, Box 24.

24. Felix Armfield, *Eugene Knickle Jones: The National Urban League and Black Social Work, 1910–1940* (Urbana: University of Illinois Press, 2012).

25. Joellen Watson Hawkins, "Adah Thoms," in *Dictionary*, ed. Kaufman, 365–367.

26. RAC, Laura Spelman Rockefeller Memorial (LSRM), Series 3.1 Box 2, Folder 22, Lincoln Hospital and Home 1920–1921.

27. CSS #0273, Box 36, Folder Columbus Hill 1923–1935, Burritt to Franklin Kirkbride, February 5, 1918.

28. MMF, Biographical Records, Group 845, Series II, Box 24.

29. CSS #0273, Box 63, Folder: Home Hospital Planning, Annual Meeting of the AICP, November 23, 1921. For nursing's role at this conference, see Cynthia Connolly, "Determining Children's Best Interests in the Middle of an Epidemic: A Cautionary Tale from History," in *Nursing Interventions Through Time: History as Evidence*, ed. Patricia D'Antonio and Sandra Lewenson (New York: Springer Publishing, 2010), 17–29.

30. Amy Fairchild, Ronald Bayer, and James Colgrove, *Searching Eyes: Privacy, the State and Disease Surveillance in America* (Berkeley: University of California Press, 2007). For children, see Cynthia Connolly, *Saving Sickly Children: The Tuberculosis Preventorium in American Life, 1909–1970* (New Brunswick, NJ: Rutgers University Press, 2008).

31. MMF, Biographical Records, Group 845, Series II, Box 24.

32. Buhler-Wilkerson, *False Dawn*.

33. Duffy, *The Sanitarians*, 214–215.

34. John Duffy, *History of Public Health in New York City, 1866–1966* (New York: Russell Sage Foundation, 1968), 272–276.

35. Editorial, "The Political Attack on the New York City Health Department," *American Journal of Public Health* 8, no. 4 (1918): 380–381.

36. Kenneth Widdemer, *A Decade of District Center Health Pioneering: East Harlem Health Center* (New York: privately published, 1932), 30.

37. Russell Leigh Sharmon, *The Tenants of East Harlem* (Berkeley: University of California Press, 2006), 24. See also Joseph Cosco, *Imagining Italians: The Clash of Romance and Race in American Perceptions* (New York: State University Press of New York, 2003), who reports that 80 percent of the 4.5 million Italian immigrants to the United States were from the south; and one-third settled in New York City.

38. Harmon, *The Tenants of East Harlem*, 25–28; Widdemer, *A Decade*, 17.

39. Susan Reverby, *East Harlem Health Center: An Anthology of Pamphlets* (New York: Garland Publishing, 1985), 17.

40. Cosco, *Imagining Italians*, 7–11.

41. See also Jennifer Gugliemo, *Living the Revolution: Italian Women's Resistance and Radicalism in New York City* (Chapel Hill: University of North Carolina Press, 2010); Thomas Gugliemo, "Encountering the Color Line in the Everyday: Italians in Interwar Chicago," *Journal of American Ethnic History* 23, no. 4 (2004): 45–77; Robert Orsi, *The Madonna of 115th Street: Faith and Community in Italian Harlem, 1880–1950* (New Haven, CT: Yale University Press, 2002); and S. M. Tomasi, *Perceptions in Italian Immigration and Ethnicity* (New York: Center for Migration Studies, 1976).

42. MMF, Group 845, Series III, Box 30, Folder 4, Study Reports, Drolet to Devine, July 10, 1930.

43. RAC, LSRM, Series 3.1, Box 1, Folder 10, 1925 Memorandum of Interview with Homer Folks.

44. Kenneth Widdemer, *The House That Health Built* (New York City: privately published, 1925).

45. RAC, LSRM, Series3.1, Box 1, Folder 12, East Harlem Health Center-Nursing, 1921.

46. RAC, LSRM. Series 3.1, Box 1, Folder 12, "Health Center to Be Established by the Red Cross and Co-operating Organizations."

47. CSS #0273, Box 63, Folder: Home Hospital Planning, October 19, 1921.

48. CSS #0273, Box 33, Folder, East Harlem Health Center, 1922–1929, "Specialized versus Generalized."

49. On "Homerian diplomacy," see CSS #0273, Box 32, Folder East Harlem Health Center, 1919–1923, Burritt to Folks, March 17, 1921. On bluff, see CSS #0273, Mullberry Street, Box 61, Folder 367–12, Gebhard to Burrett, February 15, 1923.

50. RAC, LSRM. Series 3.1, Box 1, Folder 12, "Health Center to Be Established by the Red Cross and Co-operating Organizations."

51. CSS #2073, Box 63, Folder: Home Hospital Planning, Annual Meeting of the AICP, November 23, 1921.

52. CSS #2073, Box 58, Victoria Apartments / Home Hospital; CSS #2073, Box 63, Home Hospital Planning, Annual Meeting of the AICP, November 23, 1921.

53. CSS #0273, Box 63, Folder: Home Hospital Planning, October 19, 1921.

54. MMF, Group 845, Series 1, Box 1, Folder: Annual Conference November 16, 1922. On Fox, see Joellen Watson Hawkins, "Elixabeth Gordon Fox," in *Dictionary*, ed. Kaufman, 127–129.

55. RAC, Series 3.1, Box 1, Folder 12, Correspondence March 16, 1922 to May 23, 1922.

56. CSS #0273, Box 62, Folder: 369–6, Conference to Consider a Plan for the Control of Tuberculosis, March 9, 1923.

57. See, for example, "Statement of Proposed Bellevue-Yorkville Health Demonstration," CSS #0273, Box 62, Folder: 369–6, January 30, 1924.

58. C.-E.A. Winslow, *Health on the Farm and in the Village* (New York: Macmillan, 1931), 1–2.

59. MMF Group 845, Series 1, Box 1, Folder 1: Advisory Committee November 16, 1922.

60. MMF, Group 845, Series 3, Box 30, Folder 1: Vital Statistics 1922–1924. It also contained the Milbank Memorial Fund's underused Bathhouse at 325 East 38th Street for which the fund had pledged monies for a renovation of the demonstration's headquarters.

61. CSS #0273, Box 62, Folder 371–8: Newspaper Clippings.

Chapter 2 — The Houses That Health Built

1. Homer Folks, Foreword, in Kenneth Widdemer, *The House That Health Built: A Report of the First Three Years of the East Harlem Health Center Demonstration* (New York: privately published, 1925), 7.

2. Grace Anderson, "An Urban Nursing and Health Demonstration," *American Journal of Nursing* 30, no. 12 (1930): 1531–1532.

3. Milbank Memorial Fund (MMF), Group No. 845, Series I, Box 10, Folder 74, Technical Board Committee Minutes, 1924; MMF, Record Group 845, Series I, Box 10, Folder 75, Technical Board Committee Minutes, 1926; MMF, Record Group 845, Series I, Box 11, Folder 77, Technical Board Minutes, 1926–1927.

4. John Duffy, *Public Health in New York City, 1866–1966* (New York: Russell Sage Foundation, 1968).

5. Health Sciences Library Archives and Special Collections, Columbia University Medical Center (CUMC), Maternity Center Association (MCA), MCA Guide to the Records.

6. CUMC, Guide.

7. CUMC, MCA, Box 2, Folder 8, 1924.

8. CUMC, MCA, Box 2, Folder 9, 1925.

9. Arnold Gessel, *The Pre-school Child: From the Standpoint of Public Hygiene and Education* (New York: Houghton Mifflin, 1923).

10. Jeffrey Brosco, "Weight Charts and Well Child Care: When the Pediatrician Became the Expert in Well Child Care," in *Formative Years: Children's Health in the United States, 1880–2000*, ed. Alexandra Minna Stern and Howard Markel (Ann Arbor: University of Michigan Press, 2002), 91–120.

11. Gessel's *The Pre-school Child*. He recommended one public health nurse for a population of 2,000 to meet minimum standards of child health and welfare from care of the mother during pregnancy through school age (200).

12. Widdemer, *The House That Health Built*, 26, 35–44.

13. Ibid., 12–13.

14. Isidore S. Falk, *The Costs of Medical Care: A Summary of Investigation on the Economic Aspects of the Prevention and Care of Illness* (Chicago: University of Chicago Press, 1933).

15. Widdemer, *The House that Health Built*, 35–95.

16. Ibid., 46–47.

17. Rockefeller Archive Center (RAC), Laura Spelman Rockefeller Memorial (LSRM), Series 3.1, Box 1, Folder 12, Governing Board Resolution, 11 May 1925.

18. CUMC, MCA, Box 56, Folder 1, Goodrich to Husk, August 29, 1922; for data about the agreement, see CUMC, MCA, Box 56, Folder 1, August 25, 1922.

19. CUMC, MCA Box 56, Folder 1, *The Manhattan Health Society: An Adventure on Self-Supporting Health Service for the Middle Class*, 12.

20. CUMC, MCA, Box 56, Folder 1, Director's Report, 1922.

21. RAC, LSRM, Series 3.1, Box 1, Folder 12, Letter Homer Folks to Memorial, March 4, 1924.

22. RAC, LSRM, Series 3.1, Box 1, Folder 12, Letter Homer Folks to Beardsley Ruml, May 12, 1925.

23. 100 Years: The Rockefeller Foundation, "Beardsley Ruml," Rockefeller1100.Org, last accessed January 24, 2015, http://rockefeller100.Org/biography/show/beardsley -ruml, retrieved January 24, 2015. Beardsley would later be remembered as the "father of modern social science." See, for example, Martin Bulmer and Joan Bulmer, "Philanthropy and Social Science in the 1920s: Beardsley Ruml and the Laura Spelman Rockefeller Memorial, 1922–1929," *Minerva* 19, no. 3 (1981): 347–407.

24. For the debate over authority, see CSS #0273, Box 10, Folder Home Hospital, 1921.

25. RAC, LSRM, Series 3.1, Box 1, Folder 10, Memorandum of Interview with Homer Folks, November 22, 1925.

26. RAC, LSRM, Series 3.1, Box 1, Folder 10, Interview with Lawson Purdy, January 5, 1926.

27. RAC, LSRM, Series 3.1, Box 1, Folder 10, Interview with Lillian Wald, January 5, 1926.

28. RAC, LSRM, Series 3.1, Box 1, Folder 10, Memorandum of Interview with Homer Folks, November 22, 1925.

29. RAC, LSRM, Series 3.1, Box 1, Folder 10, Interview with Mr. Burritt, March 9. 1926.

30. On the Welfare Council charge, see Duffy, *Public Health in New York City*, 306–307.

31. RAC, LSRM, Series 3.1, Box 1, Folder 10, Evolution of the East Harlem Health Center to the East Harlem District Center, January 14, 1926.

32. RAC, LSRM, Series 3.1, Box 1, Folder 10, Beardsley Ruml to Homer Folks, April 19, 1927. Other projects in healthcare the Memorial had supported included New York City's Judson Health Center and ensuring the survival of the Lincoln Hospital Training School for Nurses, acknowledged by the Memorial to be among the best training schools for black nurses.

33. RAC, RF, Record Group 1.1, Series 235, Box 1, Folder 7 Letter Homer Folks to Thomas Appleget, October 13, 1931.

34. RAC, RF, Record Group 1.1, Series 235, Box 2, Folder 16, Fifteen Years of Cooperative Endeavor: Should It Go On?

35. RAC, RF, RG 1.1, Series 235, Box 2, Folder 14: Publications, *A Comparative Study of Generalized and Specialized Nursing and Health Services*, October 1926.

36. CSS #0273, Box 31, AICP, Folder: East Harlem Health Demonstration 1934–1938, LH Gillett to Bailey Burritt, June 25, 1935.

37. "Is the Public Health Nurse a Carrier of Infection," *American Journal of Public Health* 16, no. 4 (1926): 346–351.

38. RAC, RF, RG 1.1, Series 235, Box 2, Folder 14: Publications, *The Cost of a Program of Health Activities with Special Emphasis on Public Health Nursing*, April 1926, 15.

39. Ibid., 12.

40. RAC, RF, RG 1.1, Series 235, Box 2, Folder 14: Publications. See, for example, his introduction to *East Harlem Nursing and Health Demonstration: The Cost of a Program of Health Activities with Special Emphasis on Public Health Nursing*.

41. CUMC, MCA, Box 45, Folder 12, Janet Geister, To the Report Committee, April 9, 1925.

42. RAC, RF, Record Group 1.1, Series 235, Box 1, Folder 9, undated note from Mary Beard, probably November 29, 1927.

43. CSS #0273, AICP, Box 34, Folder: East Harlem Nursing and Health Demonstration, 1920–1928, Burritt to Kingsbury, March 16, 1928.

44. RAC, RF, Record Group 1.1, Series 235, Box 2, Folder 17, *East Harlem Nursing and Health Service: A Historical Sketch* (New York: East Harlem Nursing and Health Service, September 1930).

45. See also *The Infant Service Report of the East Harlem and Nursing Health Demonstration*, 1928. Reprinted in Susan Reverby, ed., *The East Harlem Health Center Demonstration: An Anthology of Pamphlets* (New York: Garland Publishing, 1985), 5–35.

46. Grace Anderson, The East Harlem Nursing and Health Demonstration, *Public Health Nurse* 15, no. 8 (1923): 405–409; Molly Pesikoff, "In a Public Health Nursing Office: A Day in the East Harlem Nursing and Health Demonstration as Recorded by a Member of the Clerical Staff," *American Journal of Nursing* 26, no. 8 (1926): 609–611.

47. These pamphlets are in RAC, RF, RG 1.1, Series 235, Box 2, Folder 14.

48. See, for example, the meeting of February 2, 1923. CSS #0273, AICP, Box 33, Folder East Harlem Health Center 1922–1929: Minutes of Meetings. The matter of uniforms was put to the Nursing Project's staff. They voted for the HSS uniforms with a different colored coat. Annie Goodrich, however, wanted the Nursing Project's uniforms to match those of HSS. The issue had been a "difficult" one and a final decision was tabled for the time being. For examples of salaries and vacation schedules, see CSS #0273, AICP, Box 34, Folder: East Harlem Nursing and Health Demonstration 1920–1928, March 20, 1923.

49. CSS #0273, AICP, Box 34, Folder: East Harlem Nursing and Health Demonstration, 1920–1928, correspondence between Burritt and Folks, December 21, 1923 and February 3, 1925. Despite the "vexed question of fees," the Nursing Project knew its success "depends on the good will of Henry Street" so it generally acquiesced.

50. Burgess's report was not published but Randal references it in the studies on nursing time and costs in the Bellevue-Yorkville demonstration. See M. G. Randal, "Family Composition Used in the Analysis of Home Visits by Public Health Nurses," *Milbank Memorial Fund Quarterly* 15, no. 3 (1937): 275–291. Burgess also wanted to reconsider how individual versus family visits were calculated in her consultation to the Mulberry Street clinic, but recognized the topic "would be rather hot probably." See CSS #0273, Mulberry Street, Box 61, Folder 367–12: June 24, 1924.

51. RAC, LSRM, Series 3.1, Box 1, Folder 11, Original Appeal, April 1927.

52. RAC, LSRM, Series 3.1, Box 1, Folder 13, "Outline of Family Case Study."

53. RAC, LSRM, Series 3.1, Box 1, Folder 13, Committee on Continuation, March 3, 1927; Original Appeal, April 1927; Memorandum.

54. RAC, LSRM, Series 3.1, Box 1, Folder 11, Original Appeal, April 1927.

55. RAC, LSRM, Series 3.1, Box 1, Folder 13, Memorandum of Interview, April 26, 1927.

56. RAC, Rockefeller Foundation (RF), Record Group 1.1; Series 235; Box 1, Folder 9, Richard M. Pearce, Concerning East Harlem Nursing and Health Demonstration Center.

57. RAC, RF, Record Group 1.1, Series 700, Box 19, Folder 137, Report on Foundation Cooperation on Nurse Training in Europe, December 5, 1923. Elizabeth Crowell supported this position, albeit from a different perspective. As she wrote to Edwin Embree in 1923, training schools had to be associated with the clinics and teaching hospitals of medical schools so that a new generation of physicians would acquire an appreciation of good nursing from the very beginnings of their career. Elizabeth Crowell to Edwin Embree, September 5, 1923. For Crowell's work in Czechoslovakia, see Elizabeth Vicker's "Frances Elizabeth Crowell and the Politics of Nursing in Czechoslovakia after the First World War," *Nursing History Review* 7 (1999): 67–96.

58. RAC, RF, RG 1.1, Series 700, Box 19, Folder 137F, Elizabeth Crowell to George Vincent, August 27, 1922.

59. RAC, RF, Record Group 1.1, Series 700, Box 19, Folder 137, Elizabeth Crowell to Edwin Embree, November 12, 1924.

60. RAC, RF, Record Group 1.1, Series 700, Box 19, Folder 139, Elizabeth Crowell to George Vincent, September 15, 1925.

61. *Nursing and Nursing Education in the United States* (New York: Macmillan, 1923): 23–24, 138.

62. Ibid., 11, 23–24.

63. Patricia D'Antonio, *American Nursing: A History of Knowledge, Authority and the Meaning of Work* (Baltimore, MD: Johns Hopkins University Press, 2010), 60–68.

64. RAC, RF, Record Group 1.1, Series 700, Box 19, Folder 139, Elizabeth Crowell to Edwin Embree, August 19, 1925.

65. RAC, RF, Record Group 1.1, Series 700, Box 19, Folder 139, Memorandum of a Conference, September 18, 1925.

66. RAC, RF, Record Group 1.1, Series 700, Box 19, Folder 139, "Comments by Miss Goodrich and Miss Clayton," August 26, 1925.

67. RAC, RF, Record Group 1.1, Series 700, Box 19, Folder 139, "Questions by Dr. Russell and Miss Read," August 27, 1925.

68. RAC, RF, Record Group 1.1, Series 700, Box 19, Folder 139, "Memo for Dr. Vincent," August 29, 1925.

69. RAC, RF, Record Group 1.1, Series 100, Box 19, Folder 137, Richard Pearse to Elizabeth Crowell, June 3, 1927.

70. RAC, RF, Record Group 3, Series 908, Box 15, Folder 165, Mary Beard's "Nursing Needs, 1937," 12, for retrospective analysis.

71. See Elizabeth Fee, "The Welch-Rose Report: Blueprint for Public Health Education in the Americas, 1992," available at http://www.deltaomega.org/documents/WelchRose.pdf, retrieved December 27, 2015; Greer Williams, "Schools of Public Health: Their Doing and Undoing," *Milbank Memorial Fund Quarterly* 54, no. 4 (1976): 489–527.

72. RAC, RF, Record Group 3.1, Series 900, Box 19, Folder 137, Addendum, January 4, 1927.

73. See Raymond Fosdick, *The Story of the Rockefeller Foundation* (New York: Harper & Brothers, 1952), 137–138.

74. RAC, RF, Record Group 1.1, Series 100, Box 19, Folder 137, Richard Pearse to Elizabeth Crowell, July 13, 1927.

75. RAC, RF, Record Group 3.1, Series 900, Box 19, Folder 137, Addendum, January 4, 1927.

76. CSS #0271, AICP, Box 34, Folder: East Harlem Nursing and Health Demonstration, 1920–1928, Burritt to Kingsbury, March 16, 1928.

77. On "social control" see MMF, Group 845, Series I, Folder: Technical Board Minute Book, 1926–1928; for descriptions of the discussions surrounding the devolution of the goals for the Bellevue-Yorkville demonstration, see CSS #0273, Box 62 Bellevue-Yorkville.

78. MMF, Record Group 845, Series I, Box 10, Folder 74, Technical Board Committee Minutes, 1924.

79. Lillian Wald, "Amelia H. Grant," *Public Health Nurse* 20, no. 5 (1928): 213–214. According to Wald, Grant traced her nursing lineage to her aunt, Lina Rogers, Wald's friend, colleague, and the first school nurse in New York City. For more on Grant,

see Alice Howell Friedman, "Amelia Grant," in *Dictionary of American Nursing Biography*, ed. Kaufman (New York: Greenwood Press, 1988), 165–167.

80. CSS #0273, Box 63, Folder 371–8B: Minutes of the Board of Managers, December 21, 1933.

81. RAC, RF, Record Group 1.1, Series 235, Box 2, Folder 13, East Harlem Nursing and Health Service, 1937, 14.

82. RAC, RF, Record Group 1.1, Series 235, Box 2, Folder 16, "The Story of the East Harlem Nursing and Health Service," 1937. It acknowledged that "people change slowly," and that one needed a long period of time "to gain their confidence through appreciation of their experiences, their felt needs, and their objectives."

Chapter 3 — Practicing Nursing Knowledge

1. Rockefeller Archive Center (RAC), Rockefeller Foundation (RF). Record Group 1.1, Series 235, Box 1, Folder 13, The Teaching Service of the East Harlem Health Demonstration Service, April 1, 1928 to October 1, 1931. This source also reports that one black nurse from Philadelphia was at the service and would be returning to practice TB nursing.

2. RAC, RF, Record Group 1.1, Series 235, Box 1, Folder 9, Homer Folks to Thomas Appleget, October 5, 1931.

3. RAC, RF, Record Group 1.1, Series 235, Box 1, Folder 10, Homer Folks to Thomas Appleget, April 20, 1932.

4. Savel Zimand, "Five Years at Bellevue-Yorkville: An Experiment in Health Center Administration," *American Journal of Public Health* 22, no. 4 (1932): 403–409.

5. "A Comparative Study of Generalized and Specialized Nursing and Health Services," in *The East Harlem Health Center Demonstration: An Anthology of Pamphlets*, ed. Susan Reverby (New York: Garland Publishing, 1985), 34.

6. Community Service Society (CSS) #0273, Columbus Hill, Box 36, Folder: Columbus Hill 1923–1935, "Columbus Health Center Nurses Ready for Their Day's Work in the Home" (typed photo caption).

7. CSS #0273, Columbus Hill, Box 36, Folder: Columbus Hill 1923–1935, "Columbus Health Center Nurses."

8. CSS #0273, Columbus Hill, Box 36, Folder 132: Burritt to James Hubert of the New York Urban League, November 27, 1931.

9. CSS #0273, Box 31, Folder: 1928–1937 East Harlem Health Center Pamphlets, Grace Anderson, "Vital Statistics and Statistical Procedures for Public Health Nurses," 7.

10. Milbank Memorial Fund (MMF), Group 845, Series 1, Box 1, Folder 5, Speech Transcripts, November 20, 1924.

11. CSS #0273, Box 31, Folder: 1928–1937 East Harlem Health Center Pamphlets: Grace Anderson, "Vital Statistics," 19–29.

12. Isabel Hampton Robb, *Nursing: Its Principles and Practices for Hospitals and Private Use*, 2nd ed. (Cleveland: E. C. Koeckert, 1903), 100.

13. CSS #0273, Box 31, Folder: 1928–1937 East Harlem Health Center Pamphlets, Grace Anderson, "Vital Statistics," 18.

14. RAC, RF, RG 1.1, Series 235, Box 2, Folder 14: Pamphlets, Lesson Plan for Maternity Classes, May 1926.

15. T. W. Galloway, *Love and Marriage: Normal Sexual Relations* (New York: Macmillan, 1924).

16. RAC, RF, Series 235, Box 2, Folder 14: Pamphlets, Lesson Plan for Maternity Classes, May 1926, 9.

17. Vern Bullough, "Carolyn Conant Van Blarkom," in *American Nursing: A Biographical Dictionary*, ed. Vern Bullough, Olga Church, and Alice Stein (New York: Garland Publishing, 1988), 327–328.

18. Cynthia Connolly's "Nurses: The Early Twentieth-Century Tuberculosis Preventorium Movement's 'Connecting Link,'" *Nursing History Review* 10 (2002): 127–157.

19. "A Comparative Study," 37–38.

20. RAC, RF, Record Group 1.1, Series 235, Box 2, Folder 14: Pamphlets, East Harlem Health Workers Handbook on Infant Development, Care, and Training, 1932.

21. Kumaravel Rajakumar, "Vitamin D, Cod-Liver Oil, Sunlight, and Rickets: A Historical Perspective," *Pediatrics* 112, no. 3 (2003): 132–135.

22. For the AICP's work see CSS #0278, Mulberry Street, Box 58, Folder 367: Program October 1929–September 1930.

23. "A Comparative Study," 38.

24. RAC, RF, Record Group 1.1, Series 235, Box 2 Folder 14: The Preschool Service of the East Harlem Nursing and Health Demonstration, 1927.

25. RAC, RF, Record Group 1.1, Series 235, Box 2 Folder 14: The Preschool Service of the East Harlem Nursing and Health Demonstration, 1928, 7. The Preschool Service also made diphtheria immunization and smallpox vaccination a routine part of its service in 1925; and by 1927 it claimed that 72 percent of children received diphtheria immunization and 31 percent were vaccinated against smallpox. See RAC, RF, RG 1.1, Series 235, Box 2, Folder 4: East Harlem Nursing and Health Service: A Historical Sketch, 1930.

26. RAC, RF, Record Group 1.1, Series 235, Box 2, Folder 14, East Harlem Nursing and Health Service: A Progress Report, 1934.

27. Ibid.

28. Ibid.

29. RAC, RF, Record Group 1.1, Series 235, Box 2, Folder 4: Nutrition Work in East Harlem, 1930.

30. For more on Leonard Covello's background and research interests, see Simone Cinotto, "Leonard Covello, the Covello Papers, and the History of Eating Habits among Italian Immigrants in New York," *Journal of American History* (September 2004): 497–521.

31. Historical Society of Philadelphia (HSP), MSS 40, Covello Papers, Box 66, Folder 8: J. Zappula. See also Folder 5: FJ Panetta and Folder 7: Adult Education for First Generation.

32. HSP, MSS 40, Covello Papers, Box 67, Folder 7: Miscellaneous clippings.

33. HSP, MSS 40, Covello Papers, Box 68, Folder: Mutual Aid Societies.

34. HSP, MSS 40, Covello Papers, Box 66, Folder 16.

35. HSP, MSS 40, Covello Papers, Box 68, Folder 1: Mutual Aid Societies: Health.

36. HSP, MSS 40, Covello Papers, Box 66, Folder 17 (no date).

37. HSP, MSS 40, Covello Papers, Box 60, Folder 13: The Italians and Dentistry.

38. HSP, MSS 40, Covello Papers, Box 66, Folder 13: Amedeo D'Aureli.

39. HSP, MSS 40, Covello Papers, Box 94, Folder 8: Health. The anticipated new hospital was never completed. But East Harlem nurses eagerly anticipated it, as it would eliminate the need to send Italian American mothers to the Harlem Hospital for "medically supervised" deliveries.

40. HSP, MSS 40, Covello Papers, Box 68, Folder 4: Naturalization Rally, 1938.

41. HSP, MSS 40, Covello Papers, Box 68, Folder 11: 1933 American Attitudes Toward the Italians.

42. HSP, MSS 40, Covello Papers, Box 66, Folder 15: BF 1941.

43. HSP, MSS #40, Covello Papers, Box 92, Folder 2: The Italian Peg in the American Hole.

44. HSP, MSS #40, Covello Papers, Box 10, Folder 5: 27 April 1937.

45. HSP, MSS #40, Covello Papers, Box 10, Folder 17: Neighborhood Health Development, 1937.

46. Steven Schlossman, "Before Home Starts: Notes toward a History of Parent Education in America, 1897–1929," *Harvard Educational Review* 46, no. 3 (1976): 436–467; quote on page 457.

47. Ibid., 452.

48. HSP, MSS #40, Covello Papers, Box 10, Folder 25: How Can the Life of the Italian Family Be Made to Seem Normal, Rather than a Problem to the Medical Social Worker?

49. HSP, MSS #40, Covello Papers, Box 66, Folder 13.

50. HSP, MSS #40, Covello Papers, Box 66, Folder 6.

51. HSP, MSS #40, Covello Papers, Box 66, Folder 16.

52. Mary Richmond, *What Is Social Case Work* (New York: Russell Sage Foundation, 1922).

53. RAC, RF, Record Group 1.1, Series 3.1, Box 1, Folder 10, Memo, April 26, 1927.

54. For a brief biography of Cannon, see Harvard Square Library, "Ida M. Cannon: Pioneer Medical Social Worker, 1877–1960," http://www.harvardsquarelibrary.org /biographies/ida-m-cannon/, retrieved December 28, 2015.

55. Data on social workers draw heavily from Daniel J. Walkowitz, *Working with Class: Social Workers and the Politics of Middle Class Identity* (Chapel Hill: University of North Carolina Press, 1999).

56. Abraham Flexner, "Is Social Work a Profession?," Address before the National Conference on Charities and Corrections, Baltimore, May 17, 1915, 12.

57. Abraham Flexner, *Medical Education in the United States and Canada* (New York: Carnegie Foundation for the Advancement of Teaching, 1910).

58. Flexner, "Is Social Work a Profession?," 16.

59. Ibid., 13.

60. Gerald Grob, *Mental Illness in American Society, 1875–1940* (Princeton, NJ: Princeton University Press, 1983), 144–178.

61. Andrew L. Morris, *The Limits of Voluntarism: Charity and Welfare from the New Deal through the Great Society* (Cambridge: Cambridge University Press, 2009). See also Schlossman, "Before Home Starts," 459–460.

62. "Editorial: Social Work and Public Health," *American Journal of Public Health* 12, no. 8 (1922): 702.

63. Elizabeth Adamson, "Function of the Public Health Nurse," *Public Health Nurse* 26, no. 10 (1934): 542–546.

64. RAC, RF, Record Group 1.1, Series 235, Box 2, Folder 14, Sybil Pease, Report of the Consultant in Mental Hygiene and Social Work, Some of the "Highlights," 1928–1934, 19e.

65. See also Sybil Pease, "Mental Hygiene Functions of the Public Health Nurse," *Annals of the American Academy of Political and Social Science* 149 (1930): 180–183; Sybil Pease, "New Frontiers in Public Health Nursing," *Canadian Nurse* 25 (1933): 136–138.

66. RAC, RF, Record Group 1.1, Series 235, Box 2, Folder 16, The East Harlem Nursing and Health Service: Fifteen Years of a Cooperative Endeavor, 8.

67. Pease, Report of the Consultant, 20.

68. Pease, "Mental Hygiene Functions," 180–181.

69. Ibid., 182.

70. Pease, "New Frontiers," 136–138.

71. The Infant Service Report of the East Harlem Nursing and Health Demonstration Project, 1928, in *The East Harlem Health Center Demonstration*, ed. Reverby, 22.

72. Pease, "Mental Hygiene Functions," 182.

73. Pease, Report of the Consultant, 20.

74. Pease, "New Frontiers," 136–138.

75. RAC, RF, Record Group 235, Box 2, Folder 15: Grace Anderson to Mary Beard, Report of the Student Service, 1936, 4.

76. RAC, RF, Record Group 1.1, Series 235, Box 2, Folder 16, The East Harlem Nursing and Health Service, 1937, 3.

77. RAC, RF, RG 235, Box 2, Folder 15: Grace Anderson to Mary Beard, Report of the Student Service, 1936, 7.

78. Sybil Pease, "The Interview in Public Health Nursing," *Public Health Nursing* 25 (1933): 136.

79. Miriam Ames, "Public Health Nursing," *American Journal of Public Health* 18, no. 10 (1928): 1316–1317. It should be noted that relationships between medical and nonmedical social workers were equally problematic. A three-year experiment to improve cooperation and coordination among New York City's hospitals and community-based social workers failed. See CSS #0273, Box 108, Folder Mental Hygiene.

80. "Report of the Committee on Psychiatric Social Work in Public Health Nursing Agencies," *Public Health Nurse* 21, no. 11 (1929): 579–583.

81. Lois Blakey, "Relations of the Psychiatric Social Worker to the Public Health Nurse," *Public Health Nurse* 22, no. 1 (1930): 26–28.

82. Marguerite Wales, "Influences of Modern Public Health and Social Movements on Nursing Education," *Public Health Nurse* 23, no. 7 (1931): 464.

83. CSS #0278, Box 22, Folder 57–1: Whither Nursing in the AICP, February 15, 1929.

84. Helen Sweet with Rona Dougall, *Community Nursing and Primary Health Care in Twentieth-Century Britain* (New York and Oxford: Routledge Press, 2008), 20–61. In mid-nineteenth-century Liverpool, William Rathbone, in consultation with Florence Nightingale, established the first philanthropic organization to provide nurses for the sick poor in their own homes. When Queen Victoria announced in 1887 that the money raised by the women of England to celebrate her Golden Jubilee would be used to support trained nursing, the resulting Queen Victoria Jubilee Institute for Nurses (QNI) took the training of district nursing as its domain.

85. Lily Kay, "Rethinking Institutions: Philanthropy as an Historiographic Problem of Knowledge and Power," *Minerva* 35 (1997): 283–293.

86. Ellen Lagemann, *The Politics of Knowledge: The Carnegie Corporation, Philanthropy, and Public Policy* (Chicago: University of Chicago Press, 1997), 67–70.

87. RAC, RF, Record Group 1.1, Series 235, Box 1, Folder 9: Homer Folks to Thomas Appleget, February 5, 1931.

88. RAC, Laura Spelman Rockefeller Memorial (LSRM), Series 3.1, Box 1, Folder 13, Committee on Continuation, March 3, 1927.

Chapter 4 — Shuttering the Service

1. Lillian Brandt, *Am Impressionistic View of the Winter of 1930–1931 in New York City* (New York: Welfare Council of New York City, February 1932), 7–8.
2. Ibid., 58–60.
3. Ibid., 12.
4. Ibid., 62.
5. Ibid., 69–70.
6. Rockefeller Archive Center (RAC), Rockefeller Foundation (RF), Record Group 1.1, Series 235, Box 1, Folder 14, "Community Service in 1934," 27.
7. RAC, RF, Record Group 1.1, Series 235, Box 1, Folder 10, "Report of the Director," May 11, 1934.
8. RAC, RF, Record Group 1.1, Series 235, Box 2, Folder 14, "Community Service in 1934: A Summary of Services Rendered," 27.
9. Daniel Walkowitz, *Working with Class: Social Workers and the Politics of Middle Class Identity* (Chapel Hill: University of North Carolina Press, 1999), chapter 4; Andrew L. Morris, *The Limits of Voluntarism: Charity and Welfare from the New Deal through the Great Society* (Cambridge: Cambridge University Press, 2009).
10. Community Service Society (CSS) #0273, Box 184, Folder Welfare Council—the Beginnings, Federated Financing of Social Agencies, May 1935.
11. Walkowitz, *Working with Class*, 130.
12. John Duffy, *The Sanitarians: A History of American Public Health* (Chicago: University of Illinois Press, 1990), 260.
13. CSS #0273, Box 35, Folder: Early Developments in Neighborhood Health Development, 1907–1926, John Gebhart, September 26, 1923.
14. NYC DOH, H34.01, Annual Reports 1870–1949, Roll 16 1926–1936, Annual Report 1929, "Guarding the Health of Seven Million People."
15. NYC DOH, H34.01, Annual Reports 1870–1949, Roll 16 1926–1936, Annual Report 1929. See also New York City Hall Library, *District Health Development*, March 1939, 3. Wynn knew, as did the nurses at East Harlem, that public health nurses had no difficulty in persuading mothers to bring young infants to the city's Baby Health Stations; the more difficult challenge was to keep them coming after their children turned two years of age.
16. CSS #0273, Box 63, Folder 371–8J, Edward Devine, "Three Elementary Principles," April 10, 1930.
17. MMF, Group No. 845, Series III, Box 30, Folder 3: Annual Reports 1927–1930, Annual Report 1927.
18. MMF, Group No. 845, Series III, Box 30, Folder 3: Annual Reports 1927–1930, Annual Report 1932.
19. MMF, Group No. 845, Series III, Box 30, Folder 3: Annual Reports 1927–1930.
20. Michael Davis and Mary C. Jarrett, *A Health Inventory of New York City: A Study of the Volume and Distribution of Health Services in the Five Boroughs* (New York: Welfare Council of New York, 1929), xiii.
21. Ibid., 36–39.
22. Ibid., 46.
23. NYC DOH, H34.01, Annual Reports 1870–1949, Roll 16 1926–1936, Burritt to Walker, October 28, 1929.
24. RAC, RF, Record Group 1, Box 1.1, Folder 1: Cornell Health Center Reports.

25. RAC, RF, Record Group 1.1, Series 235, Box 1, Folder 11: Thomas B. Appleget's Diary, April 5, 135.

26. RAC, RF, Series 235, RG 1.1, Box 1, Folder 11: Letter March 26, 1935.

27. To trace these negotiations, see RAC, RF, Series 235, RG 1, Folder 1: Cornell Health Center Reports.

28. HSP, Covello Papers, MSS 40, Box 65, Folder 17.

29. RAC, RF, Record Group 1.1, Series 235, Box 2, Folder 14: East Harlem Nursing and Health Service Report 1934.

30. Milbank Memorial Fund (MMF), Group 845, Series III, Box 31, Folder 6: Bellevue-Yorkville Health News, Nursing Consultation for Parents in the Schools.

31. MMF, Group 845, Series III, Box 31, Folder 6: Bellevue-Yorkville Health News, February 1930.

32. RAC, RF, Record Group 1.1, Series 235, Box 2, Folder 14, East Harlem Nursing and Health Service: A Progress Report 1934.

33. RAC, RF, Record Group 1.1, Series 235, Box 1, Folder 11, "Meeting of the Board of Directors," February 28, 1935. Also the families seemed relatively wealthier. Twenty-seven percent of mothers were able to afford the private New York hospital. The hospitals primarily used include Metropolitan (29 percent), the New York Hospital (27 percent), and Harlem Hospital (14 percent).

34. Other services provided in East Harlem: vaccinations and diphtheria immunizations; only agency in area working with children between two and school age.

35. RAC, RF, Record Group 1.1, Series 235, Box 2, Folder 14, Report of the Consultant in Parent Education, 20.

36. RAC, RF, RG 1.1, Series 235, Box 2, Folder 14, Report of the Consultant in Mental Hygiene and Social Work, 1928–1934, 20.

37. RAC, RF, Record Group 1.1, Series 235, Box 2, Folder 16, The Story of the East Harlem Nursing and Health Service, 1937.

38. RAC, RF, Record Group 1.1, Series 235, Box 2, Folder 14, East Harlem Nursing and Health Services in 1934.

39. RAC, RF, Record Group 1.1, Series 235, Box 2, Folder 16, A Brief Review of Certain Aspects of the Year's Work, 1936.

40. RAC, RF, Series 235, Box 1, Folder 11: Mary Beard's Diary, 1935.

41. RAC, RF, Record Group 1.1, Series 235, Box 2, Folder 16, The Story.

42. Clara Rue, "The School of Nursing and Fundamental Needs in Public Health Nursing," *American Journal of Nursing* 32, no. 4 (1932): 421–424.

43. Mary Roberts, "Current Events and Trends in Nursing," *American Journal of Nursing* 39, no. 1 (1939): 1–8.

44. Donald Armstrong, "The Physician and the Visiting Nurse Association," *Public Health Nurse* 26, no. 11 (1934): 580. For biographical data on Armstrong, see MMF, Group 845, Series II, Box 24, Folder: Biographical Records.

45. CSS #0273, Box 122, Folder: East Harlem Health Center 1932–1935.

46. CSS #0273, Box 36, Folder: Columbus Hill 1923–1935, Columbus Hill Report 1934; see also CSS #0273, Box 36, Folder: Columbus Hill 1923–1935, Columbus Hill (no date).

47. CSS #0273, Box 122, Folder: East Harlem Health Center 1932–1935.

48 RAC, RF, RG 1.1, Series 235, Box 2, Folder 16: A Brief Review of the Year's Work: 1936. Two-thirds of the morbidity cases were children.

49. MCA, Box 1, Folder 5: Bellevue-Yorkville.

50. MCA, Box 2, Folder 2: Governing Board Meeting April 16, 1928.

51. MCA, Box 1, Folder 11: Remarks of the Commissioner of Health.
52. MCA, Box 1, Folder 15: 1931.
53. MCA, Box 1, Folders 16 and 17.
54. "White House Council on Child Health and Protection: Conclusions of the Sub-Committee on Teaching and Education of Nurses and Nurses Attendants," *American Journal of Nursing* 31, no. 5 (1931): 584.
55. Editorial, "Unnecessary Maternal Deaths," *American Journal of Nursing* 33, no. 5 (1933): 472.
56. MCA, Box 1 Folder 18: 1934.
57. RAC, RF, RG 1.1, Series 235, Box 2, Folder 15: East Harlem Nursing and Health Service Reports 1935–1936, December 1935.
58. RAC, RF, Record Group 1.1, Series 235, Box 2, Folder 16: The East Harlem Nursing and Health Service, 1937, 6.
59. RAC, RF, RG 1.1, Series 235, Box 1, Folder 11: Minutes of the Executive Board, November 12, 1935.
60. RAC, RF, Record Group 1.1, Series 235, Box 1, Folder 11, "Note to Dr. Russell," March 15, 1935.
61. And this included the Henry Street Settlement and VNS—deeply worried it had no sustainable endowment. Finally gave it a modest one and cut off yearly appropriations, a common move on the part of the Foundation and the projects it sponsored.
62. RAC, RF, Record Group 1.1, Series 235, Box 1, Folder 11, "Note to Dr. Russell," March 15, 1935.
63. RAC, RF, Record Group 1.1, Series 235, Box 1, Folder 11, "Excerpt from Thomas B. Appleget's (TBA) Diary 22 April 1935."
64. RAC, RF, Record Group 1.1, Series 235, Box 1, Folder 11, TBA to Folks, June 24, 1935.
65. RAC, RF, Record Group 1.1, Series 235, Box 1, Folder 11, Mary Beard's Diary, November 12, 1935.
66. RAC, RF, Record Group 1.1, Series 235, Box 1, Folder 11, Mary Beard, "The East Harlem Nursing and Health Service," December 11, 1935.
67. RAC, RF, Record Group 1.1, Series 235, Box 1, Folder 10, Excerpt from TBA's Diary, May 3, 1937.
68. RAC, RF, Record Group 1.1, Series 235, Box 1, Folder 12, Mary Beard to Homer Folks, January 16, 1936.
69. See, for example, "District Health Administration in New York City," *Milbank Memorial Fund Quarterly* 11, no. 3 (1933): 208–220.
70. RAC, RF, Record Group 1.1, Series 235, Box 2, Folder 16, "The East Harlem Nursing and Health Service: Fifteen Years of a Cooperative Endeavor: Should It Go On?" 1937.
71. RAC, RF, Record Group 1.1, Series 235, Box 2, Folder 13, The East Harlem Nursing and Health Service, 1937.
72. RAC, RF, Record Group 1.1, Series 235, Box 2, Folder 16, Katherine Tucker to Grace Anderson, March 10, 1937.
73. RAC, RF, Record Group 1.1, Series 235, Box 2, Folder 16, Marguerite Wales to Grace Anderson, March 19, 1937. Wales mentioned a recent Rockefeller Foundation–funded trip to Central Europe where former foundation fellows eagerly read East Harlem reports. Other letters of support came from Elizabeth Gordon Fox, executive director of the Visiting Nurse Association of New Haven (RAC, RF, Record Group 1.1, Series 235, Box 2, Folder 16, Fox to Anderson, April 1, 1937); Lillian Hudson

of Teachers College (RAC, RF, Record Group 1.1, Series 235, Box 2, Folder 16, April 12, 1937); and Amelia Grant, director of the Bureau of Nursing in the Department of Health (RAC, RF, Record Group 1.1, Series 235, Box 2 Folder 16, Grant to Anderson, May 10, 1937).

74. RAC, RF, Record Group 1.1, Series 235, Box 2, Folder 16, Katherine Tucker to Grace Anderson, March 10, 1937.

75. RAC, RF, RG 1.1 Series 235, Box 1, Folder 1: Tucker to Anderson December 12, 1937. For data on Tucker, see Signe Cooper, "Katherine Ellen Faville," in *American Nursing: A Biographical Dictionary*, ed. Vern Bullough, Lilli Sentz, and Alice Stein, vol. 2 (New York: Garland Publishing, 1992), 108–110.

76. MCA, Box 2, Folder 3.

77. RAC, RF, RG 1.1, Series 235, Box 1, Folder 12, East Harlem Nursing and Health Service 1936–1941.

78. MCA, Box 2, Folder 6.

79. RAC, RF, Record Group 1.1, Series 235, Box 1, Folder 12, Margaret Nourse to Raymond Fosdick and TBA, January 15, 1941.

Chapter 5 — Not Enough to Be a Messenger

1. "Presidential Proclamation—National Health Center Week, 2015," https://www .whitehouse.gov/the-press-office/2015/08/07/presidential-proclamation-national -health-center-week-2015. Retrieved September 4, 2015.

2. See, for example, American Academy for Oral Systemic Health, http://aaosh.org. Retrieved September 4, 2015.

3. Michael Davis, "The Voluntary Agency in a Democracy," *Public Health Nurse* 31, no. 2 (1939): 193.

4. Karen Buhler-Wilkerson, *False Dawn: The Rise and Fall of Public Health Nursing in the United States, 1900–1930* (New York: Garland Press, 1990).

5. Lorin Reed Thomas, "Citizens in the Margins: Puerto Rican Migrants in New York City, 1917–1960" (PhD diss., University of Pennsylvania, 2002); Virginia Sánchez Korral, *From Colonia to Community: The History of Puerto Ricans in New York City* (Berkeley: University of California Press, 1994).

6. "The Harlem Renaissance," http://www.literaryhistory.com/20thC/HarlemRen.htm. Retrieved March 7, 2015.

7. Rosemary Stevens, *In Sickness and in Wealth: American Hospitals in the Twentieth Century* (Baltimore, MD: Johns Hopkins University Press, 1989), 106.

8. Ibid., chapter 6 and p. 143.

9. Amy L. Fairchild et al., "The Exodus of Public Health: What History Can Tell Us About the Future," *American Journal of Public Health* 100, no. 1 (2010): 54–63.

10. John Kingsbury, the executive director of the Milbank Memorial Fund, was one striking casualty of the prevailing conservatism. Angered that the 1932 Committee on the Costs of Medical Care, a national study of ability (or, more correctly, inability) of American families to cover the costs of health and illness care, stopped short of recommending federal financing, he pursued his own advocacy. What the fund believed to be his overstepping of boundaries between its policies and political advocacy ultimately led to a national rebuke by Alfred Milbank and, shortly thereafter, Kingsbury's loss of his position. See Daniel Fox, "The Significance of the Milbank Memorial Fund for Policy: An Assessment at Its Centenary," *Milbank Quarterly* 84, no. 1 (2006): 5–36.

11. Patricia D'Antonio, *American Knowledge: A History of Power, Authority and the Meaning of Work* (Baltimore, MD: Johns Hopkins University Press, 2010).

12. Patricia D'Antonio, Linda Beeber, Grayce Sills, and Madeline Naegle, "The Future in the Past: Hildegard Pelpau and Interpersonal Relationships in Nursing," *Nursing Inquiry* 21, no. 4 (2014): 311–317.

13. J. C. Hart, "As Others See Us," *Public Health Nurse* 21, no. 7 (1929): 343–345.

14. Savel Zimand, "Five Years at Bellevue-Yorkville: An Experiment in Health Center Administration," *American Journal of Public Health* 221, no. 4 (1932): 403–409.

15. MMF, Group 845, Series 1, Box 11, Folder: Technical Board Minute Books 1932–1935, December 6, 1934.

16. Marjorie Feld, *Lillian Wald: A Biography* (Chapel Hill: University of North Carolina Press, 208), 175.

17. Historical Society of Pennsylvania, MSS 40, Covello Papers, Box 71, Folder 1.

18. Karen Buhler Wilkerson, *No Place Like Home: A History of Nursing and Home Care in the United States* (Baltimore, MD: Johns Hopkins University Press, 2003), 712.

19. Julio Frenk et al., "Health Professionals for a New Century: Transforming Education to Strengthen Health Systems in an Interdependent World," *Lancet* 376, no. 9756 (2010): 1923–1958.

20. Institute of Medicine and the Committee on the Robert Wood Johnson Foundation Initiative on the Future of Nursing at the Institute of Medicine, *The Future of Nursing: Leading Change, Advancing Health* (Washington, DC: National Academies Press, 2010).

Bibliography

Apple, Rima. *Mothers and Motherhood: A Social History of Infant Feeding, 1890–1950.* Madison: University of Wisconsin Press, 1987.

———. *Perfect Motherhood: Science and Childrearing in America.* New Brunswick, NJ: Rutgers University Press, 2006.

Armfield, Felix. *Eugene Knickle Jones: The National Urban League and Black Social Work, 1910–1940.* Urbana: University of Chicago Press, 2012.

Armstrong, Donald. "The Physician and the Visiting Nurse Association." *Public Health Nurse* 26, no. 11 (1934): 578–581.

Beatty, Barbara, Emily Cahan, and Julie Grant, eds. *When Science Encounters the Child: Education, Parenting, and Child Welfare in Twentieth-Century America.* New York: Teachers College Press, 2006.

Brandt, Lillian. *An Impressionistic View of the Winter of 1930–1931 in New York City.* New York: Welfare Council of New York City, February 1932.

Brosco, Jeffrey. "Weight Charts and Well Child Care: When the Pediatrician Became the Expert in Well Child Care." In *Formative Years: Children's Health in the United States, 1880–2000,* edited by Alexandra Minna Stern and Howard Markel, 91–120. Ann Arbor: University of Michigan Press, 2002.

Buhler-Wilkerson, Karen. *False Dawn: The Rise and Decline of Public Health Nursing in the United States, 1900–1930.* New York: Garland Press, 1989.

———. *No Place Like Home: Nursing and Home Care in the United States.* Baltimore, MD: Johns Hopkins University Press, 2003.

Bullough, Vern, Olga Church, and Alice Stein, eds. *American Nursing: A Biographical Dictionary.* New York: Garland Publishing, 1988.

Bullough, Vern, Lilli Sentz, and Alice Stein, eds. *American Nursing: A Biographical Dictionary,* vol. 2. New York: Garland Publishing, 1992.

Cinotto, Simone. "Leonard Covello, the Covello Papers, and the History of Eating Habits among Italian Immigrants in New York." *Journal of American History* (September 2004): 497–521.

Colgrove, James. *State of Immunity: The Politics of Vaccination in Twentieth-Century America.* Berkeley: University of California Press and the Milbank Memorial Fund, 2006.

Connolly, Cynthia. "Nurses: The Early Twentieth Century Tuberculosis Preventorium Movement's 'Connecting Link.'" *Nursing History Review* 10 (2002): 127–157.

———. *Saving Sickly Children: The Tuberculosis Preventorium in American Life, 1909–1970.* New Brunswick, NJ: Rutgers University Press, 2008.

Cosco, Joseph. *Imagining Italians: The Clash of Romance and Race in American Perceptions.* New York: State University Press of New York, 2003.

D'Antonio, Patricia. *American Nursing: A History of Knowledge, Authority, and the Meaning of Work.* Baltimore, MD: Johns Hopkins University Press, 2010.

D'Antonio, Patricia, Linda Beeber, Grayce Sills, and Madeline Naegle. "The Future in the Past: Hildegard Pelpau and Interpersonal Relationships in Nursing." *Nursing Inquiry* 21, no. 4 (2014): 311–317.

Davis, Michael. "The Voluntary Agency in a Democracy." *Public Health Nurse* 31, no. 2 (1939): 191–194.

Davis, Michael, and Mary C. Jarrett. *A Health Inventory of New York City: A Study of the Volume and Distribution of Health Services in the Five Boroughs.* New York: Welfare Council of New York, 1929.

District Health Administration in New York City. *Milbank Memorial Fund Quarterly* 11, no. 3 (1933): 208–220.

Dixon Vuic, Kara. "Wartime Nursing and Power." In *Routledge Handbook on the Global History of Nursing,* edited by Patricia D'Antonio, Julie Fairman, and Jean Whelan, 22–34. Oxford: Routledge, 2014.

Duffy, John. *History of Public Health in New York City, 1866–1966.* New York: Russell Sage Foundation, 1968.

———. *The Sanitarians: A History of American Public Health.* Chicago: University of Illinois Press, 1990.

Editorial, "Unnecessary Maternal Deaths." *American Journal of Nursing* 33, no. 5 (1933): 472.

Fairchild, Amy, Ronald Bayer, and James Colgrove. *Searching Eyes: Privacy, the State, and Disease Surveillance in America.* Berkeley: University of California Press, 2007.

Fairchild, Amy, David Rosner, James Colgrove, Ronald Bayer, and Linda Fried. "The Exodus of Public Health: What History Can Tell Us about the Future." *American Journal of Public Health* 100, no. 1 (2010): 54–63.

Farley, John. *To Cast Out Disease: A History of the International Health Division of the Rockefeller Foundation, 1913–1951.* Oxford and New York: Oxford University Press, 2004.

Feld, Marjorie N. *Lillian Wald: A Biography.* Chapel Hill: University of North Carolina Press, 2008.

Flexner, Abraham. "Is Social Work a Profession?" Address before the National Conference on Charities and Corrections, Baltimore, May 17, 1915.

———. *Medical Education in the United States and Canada.* New York: Carnegie Foundation for the Advancement of Teaching, 1910.

Fosdick, Raymond. *The Story of the Rockefeller Foundation.* New York: Harper & Brothers, 1952.

Fox, Daniel M. "The Significance of the Milbank Memorial Fund for Policy: An Assessment at Its Centennial." *Milbank Quarterly* 84, no. 1 (2006): 5–36.

Frenk, Julio, Lincoln Chen, Zulfiqar Bhutto, Jordan Cohen, Nigel Crisp, Timothy Evans, Harvey Fineberg, Patricia Garcia, Yang Ke, Patrick Kelley, Barry Kistnasamy, Afaf Meleis, David Naylor, Ariel Pablos-Mendez, Srinath Reddy, Susan Scrimshaw, Jamie Sepilveda, David Serwadda, and Hilda Zurayk. "Health Professionals for a New Century: Transforming Education to Strengthen Health Systems in an Interdependent World." *The Lancet* 376, no. 9756 (2010): 1923–1958.

Gessel, Arnold. *The Pre-school Child: From the Standpoint of Public Hygiene and Education.* New York: Houghton Mifflin Company, 1923.

Golden, Janet. *A Social History of Wet Nursing: From Breast to Bottle.* Cambridge: Cambridge University Press, 1990.

Grant, Julie. *Raising Baby by the Book: The Education of Modern Motherhood.* New Haven, CT: Yale University Press, 1998.

Grob, Gerald. *Mental Illness in American Society, 1875–1940.* Princeton, NJ: Princeton University Press, 1983.

Gugliemo, Jennifer. *Living the Revolution: Italian Women's Resistance and Radicalism in New York City.* Chapel Hill: University of North Carolina Press, 2010.

Gugliemo, Thomas. "Encountering the Color Line in the Everyday: Italians in Interwar Chicago." *Journal of American Ethnic History* 23, no. 4 (2004): 45–77.

Hart, J. C. "As Others See Us." *Public Health Nurse* 21, no. 7 (1929): 343–345.

Institute of Medicine and the Committee on the Robert Wood Johnson Foundation Initiative on the Future of Nursing at the Institute of Medicine. *The Future of Nursing: Leading Change, Advancing Health.* Washington, DC: National Academies Press, 2010.

Jones, Kathleen W. *Taming the Troublesome Child: American Families, Child Guidance, and the Limits of Psychiatric Authority.* Cambridge, MA: Harvard University Press, 1999.

Kaufman, Martin, ed. *Dictionary of American Nursing Biography.* New York: Greenwood Press, 1988.

Kay, Lily. "Rethinking Institutions: Philanthropy as an Historiographic Problem of Knowledge and Power." *Minerva* 35 (1997): 283–293.

Kohler, Robert. *Foundations and Natural Scientists, 1900–1945.* Chicago: University of Chicago Press, 1981.

Korral, Virgina Sánchez. *From Colonia to Community: The History of Puerto Ricans in New York City.* Berkeley: University of California Press, 1994.

Ladd Taylor, Molly. *Motherwork: Women, Child Welfare and the State 1890–1930.* Urbana and Chicago: University of Illinois Press, 1994.

Lagemann, Ellen. *Politics of Knowledge: The Carnegie Corporation, Philanthropy, and Public Policy.* Middleton, CT: Wesleyan University Press, 1989.

Ludmerer, Kenneth. *Learning to Heal: The Development of American Medical Education.* New York: Basic Books, 1985.

Morris, Andrew L. *The Limits of Voluntarism: Charity and Welfare from the New Deal through the Great Society.* Cambridge and New York: Cambridge University Press, 2009.

Orsi, Robert. *The Madonna of 115th Street: Faith and Community in Italian Harlem, 1880–1950.* New Haven, CT: Yale University Press, 2002.

Picard, Alyssa. *Making the American Mouth: Dentists and Public Health in the Twentieth Century.* New Brunswick, NJ: Rutgers University Press, 2009.

Reverby, Susan. *East Harlem Health Center: An Anthology of Pamphlets.* New York: Garland Publishing, 1985.

———. *Ordered to Care: The Dilemma of American Nursing, 1850–1945.* Cambridge: Cambridge University Press, 1987.

Roberts, Mary. "Current Events and Trends in Nursing." *American Journal of Nursing* 39, no. 1 (1939): 1–8.

Rosen, George. "The First Neighborhood Health Center Movement: Its Rise and Fall." *American Journal of Public Health* 61 (1971): 1620–1635.

Rue, Clara. "The School of Nursing and Fundamental Needs in Public Health Nursing." *American Journal of Nursing* 32, no. 4 (1932): 421–425.

Sharmon, Russell Leigh. *The Tenants of East Harlem.* Berkeley: University of California Press, 2006.

Stevens, Rosemary. *In Sickness and in Wealth: American Hospitals in the Twentieth Century.* Baltimore, MD: Johns Hopkins University Press, 1989.

Sweet, Helen, with Rona Dougal. *Community Nursing and Primary Health Care in Twentieth-Century Britain.* New York and Oxford: Routledge, 2008.

Thomas, Lorin Reed. "Citizens in the Margins: Puerto Rican Migrants in New York City, 1917–1960." PhD diss., University of Pennsylvania, 2002.

Tomasi, S. M. *Perceptions in Italian Immigration and Ethnicity.* New York: Center for Migration Studies, 1976.

Toon, Elizabeth. "Selling the Public on Health: The Commonwealth and Milbank Health Demonstrations and the Meaning of Community Health Education." In *Philanthropic*

Foundations: New Scholarship, New Possibilities, edited by Ellen Lagemann, 119–130. Bloomington: Indiana University Press, 1999.

Vandenberg-Daves, Jodi. *Modern Motherhood: An American History.* New Brunswick, NJ: Rutgers University Press, 2014.

Vicker, Elizabeth. "Frances Elizabeth Crowell and the Politics of Nursing in Czechoslovakia after the First World War." *Nursing History Review* 7 (1999): 67–96.

Walkowitz, Daniel. *Working with Class: Social Workers and the Politics of Middle Class Identity.* Chapel Hill: University of North Carolina Press, 1999.

Wardo, John, and Warren, Christopher, eds. *Silent Victories: The History and Practice of Public Health in Twentieth-Century America.* Oxford and New York: Oxford University Press, 2007.

White House Council on Child Health and Protection. "Conclusions of the Sub-Committee on Teaching and Education of Nurses and Nurses Attendants." *American Journal of Nursing* 31, no 5 (1931): 584–586.

Widdemer, Kenneth. *A Decade of District Center Health Pioneering: East Harlem Health Center.* New York: privately published, 1932.

———. *The House That Health Built.* New York: privately published, 1925.

Williams, Greer. "Schools of Public Health: Their Doing and Undoing." *Milbank Memorial Fund Quarterly* 54, no. 4 (1976): 489–527.

Wolfe, Jacqueline H. *Deliver Me from Pain: Anesthesia and Birth in America.* Baltimore, MD: Johns Hopkins University Press, 2009.

Zimand, Savel. "Five Years at Bellevue-Yorkville: An Experiment in Health Center Administration." *American Journal of Public Health* 22, no. 4 (1932): 403–409.

Index

Page references followed by f indicate a figure, by t a table.

psychiatric issues, in First World War's draft screenings, 71. *See also* mental health; mental hygiene
psychoanalytical theory, 71
public health: global programs in, 48–49; initiatives, 101; medicalized model of, 108; new science of, 104; postgraduate education in, 50; public and private interests in, 36, 38; Rockefeller Foundation's support of, 48
public healthcare: hierarchical structure of, 110; responsibility for setting agendas for, 111
public health disciplines, core of, 112
public health nurses: black, 59; and care coordination, 14; central role of, 2–3; changing roles of, 62; data collected by, 61; in demonstration projects, 4; and the Depression, 80; education of, 57–58, 91; in England, 76; Flexner's views on, 71; graduate preparation of, 53; and home hospital concept, 31; under Mayor LaGuardia, 82–83; nursing supervision of, 38; in NYC, 9–10, 88; NYC's proposed coordination of, 29; private, 9; progressive, 109; social casework methods used by, 72; and social workers, 70–76, 72; on subsidies for poor, 109; success of, 87; supervision of, 10, 108; and TB eradication, 13; and TB reporting and monitoring, 24–25; traditional checklists of, 69–70; trained in East Harlem, 97
public health nursing: classroom content for, 19; curricular leadership in, 77; and demonstration projects, 103; early training for, 18–19; educational advancement of, 11; education for, 49, 50; higher education in, 49; histories of, 8; impact of the Depression on, 79, 81; mental hygiene in, 71–72, 74, 90; modern, 75–76; "new approach" in, 90; in NYC, 86–88; postgraduate programs in, 45; public *vs.* private, 2; specialization in, 30; support for, 7; teaching in, 45
Public Health Nursing Bureau, NYC's, 65
public health nursing education: lack of support for, 54; postgraduate, 95; teaching field for, 95

public health nursing leadership: discrimination of, 23; and East Harlem Nursing Project, 41
public health nursing practice, 58; biological sciences in, 59; black nurses in, 59–60; control of, 88; education for, 18; European models of, 50; family nursing practice, 64–65; and generalization *vs.* specialization, 30; generalized, 63; impact of school on, 88; insularity of, 109; in interwar years, 110–111; knowledge for, 59–64; "new approach" to, 72, 74; physicians in, 105; planning for, 38; preschool children in, 63–64, 64f; and problems of coordination, 59; recording keeping in, 61
public health nursing reformers, 14; M. Beard, 20–21; L. Clayton, 20; J. Goldmark, 20; A. Goodrich, 19; and nursing goals, 4; M. A. Nutting, 19–20; and public health nursing standards, 15; L. Wald, 19
public health nursing students, costs of teaching, 45
public health promotion, and private medical practice, 85
Puerto Rican families, 80; in East Harlem, 10; and neighborhood health centers, 102–103; and public health nurses, 107–108
Puerto Rican migrants, 9
Purdy, Lawson, 43

quality improvement data, 2
quarantine, in early public health nursing, 18
Queens, NYC, health districts of, 26
Queen Victoria Jubilee Institute for Nurses (QNI), 124n84

race: at Columbus Hill Health Center, 60; at East Harlem Nursing Service, 60; and health, 106
race riot, first modern, 103
Rathbone, William, 124n84
registered nurse, title of, 18
Richmond, Mary E., 70–71
rickets, prevention of, 63
Robb, Isabel Hampton, 61

About the Author

Patricia D'Antonio is the Killebrew-Censits Endowed Term Professor of Undergraduate Education, chair of the Department of Family and Community Health, and director of the Barbara Bates Center for the Study of the History of Nursing at the University of Pennsylvania School of Nursing. She is a senior fellow of the Leonard Davis Institute of Health Economics and a member of the core faculty of the Alice Paul Center for Research on Women, Gender, and Sexuality at the University. D'Antonio is editor of the *Nursing History Review*, the official journal of the American Association for the History of Nursing, and author of, most recently, *American Nursing: A History of Knowledge, Authority, and the Meaning of Work.*

Available titles in the Critical Issues in Health and Medicine series:

Mark A. Hall and Sara Rosenbaum, eds., *The Health Care "Safety Net" in a Post-Reform World*

Laura L. Heinemann, *Transplanting Care: Shifting Commitments in Health and Care in the United States*

Laura D. Hirshbein, *American Melancholy: Constructions of Depression in the Twentieth Century*

Laura D. Hirshbein, *Smoking Privileges: Psychiatry, the Mentally Ill, and the Tobacco Industry in America*

Timothy Hoff, *Practice under Pressure: Primary Care Physicians and Their Medicine in the Twenty-first Century*

Beatrix Hoffman, Nancy Tomes, Rachel N. Grob, and Mark Schlesinger, eds., *Patients as Policy Actors*

Ruth Horowitz, *Deciding the Public Interest: Medical Licensing and Discipline*

Rebecca M. Kluchin, *Fit to Be Tied: Sterilization and Reproductive Rights in America, 1950–1980*

Jennifer Lisa Koslow, *Cultivating Health: Los Angeles Women and Public Health Reform*

Susan C. Lawrence, *Privacy and the Past: Research, Law, Archives, Ethics*

Bonnie Lefkowitz, *Community Health Centers: A Movement and the People Who Made It Happen*

Ellen Leopold, *Under the Radar: Cancer and the Cold War*

Barbara L. Ley, *From Pink to Green: Disease Prevention and the Environmental Breast Cancer Movement*

Sonja Mackenzie, *Structural Intimacies: Sexual Stories in the Black AIDS Epidemic*

David Mechanic, *The Truth about Health Care: Why Reform Is Not Working in America*

Richard A. Meckel, *Classrooms and Clinics: Urban Schools and the Protection and Promotion of Child Health, 1870–1930*

Alyssa Picard, *Making the American Mouth: Dentists and Public Health in the Twentieth Century*

Heather Munro Prescott, *The Morning After: A History of Emergency Contraception in the United States*

James A. Schafer Jr., *The Business of Private Medical Practice: Doctors, Specialization, and Urban Change in Philadelphia, 1900–1940*

David G. Schuster, *Neurasthenic Nation: America's Search for Health, Happiness, and Comfort, 1869–1920*

Karen Seccombe and Kim A. Hoffman, *Just Don't Get Sick: Access to Health Care in the Aftermath of Welfare Reform*

Leo B. Slater, *War and Disease: Biomedical Research on Malaria in the Twentieth Century*

Paige Hall Smith, Bernice L. Hausman, and Miriam Labbok, *Beyond Health, Beyond Choice: Breastfeeding Constraints and Realities*

Matthew Smith, *An Alternative History of Hyperactivity: Food Additives and the Feingold Diet*

Rosemary A. Stevens, Charles E. Rosenberg, and Lawton R. Burns, eds., *History and Health Policy in the United States: Putting the Past Back In*

Barbra Mann Wall, *American Catholic Hospitals: A Century of Changing Markets and Missions*

Frances Ward, *The Door of Last Resort: Memoirs of a Nurse Practitioner*

Printed and bound by CPI Group (UK) Ltd, Croydon, CR0 4YY

27/10/2024

14580231-0004